Physical Chemistry

The Comprehensive Reference

Steven Johnson

© 2024 by Wang Press. All rights reserved.

No part of this publication may be reproduced, distributed, or transmitted in any form or by any means, including photocopying, recording, or other electronic or mechanical methods, without the prior written permission of the publisher, except in the case of brief quotations embodied in critical reviews and certain other noncommercial uses permitted by copyright law.

Published by Wang Press

For permissions and other inquiries, write to:
P.O. Box 3132, Framingham, MA 01701, USA

Contents

1 Introduction to Physical Chemistry — 11
    1.1 Definition and Scope of Physical Chemistry . . . . — 11
    1.2 Historical Development and Theories . . . . — 14
    1.3 Mathematical Foundations . . . . — 18
    1.4 Units, Measurements, and Error Analysis . . . . — 21
    1.5 Interdisciplinary Connections . . . . — 24
    1.6 Applications and Importance . . . . — 27

2 Quantum Chemistry and Atomic Structure — 31
    2.1 Fundamentals of Quantum Mechanics . . . . — 31
    2.2 The Schrödinger Equation . . . . — 34
    2.3 Quantum Mechanical Models of the Atom . . . . — 38
    2.4 Atomic Orbitals and Electron Configurations . . . . — 40
    2.5 Quantum Numbers and Atomic Structure . . . . — 43
    2.6 Approximation Methods in Quantum Chemistry . . . . — 46
    2.7 Applications of Quantum Chemistry in Spectroscopy . . . . — 49

3 Molecular Bonding and Structure — 53
    3.1 Types of Chemical Bonds . . . . — 53

3.2	Valence Bond Theory	56
3.3	Molecular Orbital Theory	59
3.4	Hybridization and Molecular Geometry	63
3.5	Intermolecular Forces	67
3.6	Polarity and Its Effects on Properties	71
3.7	Resonance and Delocalized Electrons	75

4 Thermodynamics and Equilibrium — **81**

4.1	Fundamental Concepts of Thermodynamics	81
4.2	The Laws of Thermodynamics	84
4.3	Enthalpy and Heat Capacity	87
4.4	Entropy and the Second Law	90
4.5	Free Energy and Chemical Potential	94
4.6	Chemical Equilibrium and Le Chatelier's Principle	97
4.7	Phase Equilibria and Phase Diagrams	101

5 Kinetics and Reaction Dynamics — **105**

5.1	Basic Principles of Chemical Kinetics	105
5.2	Rate Laws and Reaction Order	110
5.3	Temperature and Reaction Rate: The Arrhenius Equation	114
5.4	Reaction Mechanisms and the Steady-State Approximation	118
5.5	Catalysis and Enzyme Kinetics	123
5.6	Transition State Theory and Reaction Dynamics	128
5.7	Collision Theory and Molecular Dynamics Simulations	132

6 Statistical Mechanics and Thermodynamic Models — **137**

6.1	Fundamentals of Statistical Mechanics	137
6.2	Boltzmann Distribution and Molecular Energies	140
6.3	Partition Functions and their Applications	143

	6.4	Classical and Quantum Statistics	147
	6.5	Thermodynamic Ensembles: Canonical, Microcanonical, and Grand Canonical	151
	6.6	Connection between Statistical Mechanics and Thermodynamics	154
	6.7	Models and Approximations in Thermodynamics	158

7 Spectroscopy and Photochemistry — 163

    7.1 Principles of Spectroscopy 163
    7.2 Infrared and Raman Spectroscopy 166
    7.3 Ultraviolet-Visible (UV-Vis) Spectroscopy 169
    7.4 Nuclear Magnetic Resonance (NMR) Spectroscopy 173
    7.5 Mass Spectrometry 176
    7.6 Fluorescence and Phosphorescence 180
    7.7 Photochemistry and Photophysical Processes 183

8 Surface Chemistry and Catalysis — 187

    8.1 Fundamentals of Surface Chemistry 187
    8.2 Adsorption Isotherms 190
    8.3 Surface Area and Porosity 192
    8.4 Catalysis and Catalytic Mechanisms 195
    8.5 Enzyme Catalysis and Industrial Applications 197
    8.6 Surface Characterization Techniques 200
    8.7 Applications of Surface Chemistry in Nanotechnology . . . 203

9 Electrochemistry and Conductivity — 207

    9.1 Fundamentals of Electrochemistry 207
    9.2 Nernst Equation and Electrochemical Potentials 210
    9.3 Galvanic and Electrolytic Cells 213

9.4	Conductivity and its Measurement	216
9.5	Applications of Electrochemistry	219
9.6	Electrochemical Techniques and Instrumentation	222
9.7	Ionic Conductivity and Transport Phenomena	226

10 Materials Chemistry and Nanotechnology — 231

10.1	Principles of Materials Chemistry	231
10.2	Types of Materials: Metals, Ceramics, Polymers, and Composites	235
10.3	Structure-Property Relationships	237
10.4	Synthesis and Fabrication of Nanomaterials	240
10.5	Characterization Techniques for Nanomaterials	243
10.6	Applications of Nanotechnology in Industry	247
10.7	Environmental and Ethical Implications of Nanotechnology	250

11 Biochemical Thermodynamics — 253

11.1	Principles of Biochemical Thermodynamics	253
11.2	Free Energy in Biological Reactions	256
11.3	Enzyme Thermodynamics and Kinetics	259
11.4	Thermodynamics of Metabolic Pathways	262
11.5	Binding and Allosteric Mechanisms	265
11.6	Thermodynamics of Membrane Transport	268
11.7	Applications of Thermodynamics in Biotechnology	271

12 Advanced Topics in Physical Chemistry — 275

12.1	Computational Chemistry and Simulation Techniques	275
12.2	Advanced Quantum Chemistry Methods	279
12.3	Nonequilibrium Thermodynamics	281
12.4	Soft Matter and Complex Fluids	285

12.5 Molecular Reaction Dynamics 287
12.6 Supramolecular Chemistry 290
12.7 Emerging Topics and Future Directions 293

Preface

In the vast expanse of scientific inquiry, a perennial question that has captivated the minds of physicists and chemists alike is: How do the fundamental interactions of subatomic particles give rise to the rich tapestry of observable chemical phenomena? This question is not merely academic; it forms the cornerstone of our text, "Physical Chemistry: The Comprehensive Reference," which seeks to unravel the complexities of chemical interactions and transformations.

The primary objective of this book is to provide a substantive and systematic exposition of physical chemistry, encompassing its core principles and advanced topics. Within these pages, we embark on a journey through the quantum realms that describe atomic and molecular structures, explore the energies that drive chemical processes in thermodynamics and kinetics, and delve into the extraordinary capabilities of spectroscopy and materials chemistry. Each chapter is meticulously crafted to build upon foundational concepts, ensuring a cohesive understanding that advances progressively to more sophisticated scientific paradigms.

Our approach is distinctive in its fusion of theoretical rigor with practical applications. Readers will find clear, detailed explanations of intricate topics complemented by a wealth of examples, illustrations, and exercises designed to fortify their understanding and engage their curiosity. The value of this text lies not only in its comprehensive content but also in its ability to foster critical thinking and problem-solving skills, essential traits for aspiring scientists, engineers, and industry professionals.

By delving into cutting-edge areas such as nanotechnology and biochemical thermodynamics, the book offers insights into the dynamic and ever-evolving landscape of physical chemistry. These topics underscore the relevance and impact of chemical principles in addressing contemporary challenges and innovation in technology and healthcare.

We invite you, dear reader, to immerse yourself in the pages of this textbook and uncover the scientific insights it holds. Whether you are a student beginning your academic pursuit in chemistry or a seasoned practitioner seeking to refresh and deepen your knowledge, this text serves as a gateway to a deeper appreciation and understanding of the physical universe. As you navigate through the chapters, may you find inspiration and clarity in the profound beauty of chemical science. Welcome to the fascinating world of Physical Chemistry.

Chapter 1

Introduction to Physical Chemistry

Physical chemistry stands at the intersection of chemistry and physics, offering a quantitative understanding of the molecular and atomic properties that govern chemical reactions and physical transformations. It elucidates the mechanisms driving phenomena at the microscopic level through universal principles such as thermodynamics, quantum mechanics, and statistical mechanics. This chapter embarks on a journey through the foundational theories and historical milestones that have sculpted the field, underscoring the synergy between experimental measurements and theoretical models. By exploring the mathematical underpinnings and interdisciplinary connections, readers will garner a holistic view of how physical chemistry serves as a cornerstone in scientific discovery and technological advancement.

1.1 Definition and Scope of Physical Chemistry

Physical chemistry serves as the bridge between macroscopic and molecular-level phenomena, providing a quantitative framework to understand the nature of chemical systems. It applies principles of physics to study chemical systems to unravel the complexities of structures, transformations, and energetic exchanges. In doing so, it focuses on defining the scope and interconnections between thermodynamics, quantum mechanics, and statistical mechanics, all fundamental domains that elucidate chemical phenomena.

The domain of physical chemistry outlines an expansive terrain where the physical laws of nature compel chemical events and transformations. Its defining feature is the use of mathematical models to capture and predict the behavior of chemical substances and mixtures. Through such models, physical chemistry aims to derive conclusions about bodily changes and reactivity patterns present in a diverse array of substances, including gases, liquids, solids, and solutions.

Thermodynamics provides an expansive framework for understanding energy changes during chemical reactions, delineating the concepts of energy conservation, entropy, and enthalpy. Physical chemistry leverages thermodynamics to describe the state functions and conditions under which a chemical system remains in equilibrium. It articulates how energy, in various forms, transitions within closed and open systems, governed by the laws of thermodynamics. This branch enables the precise prediction of reaction spontaneity and directionality based on Gibbs free energy, enhancing the understanding of equilibrium and non-equilibrium states.

Quantum chemistry extends the physical chemistry realm into the subatomic domain, where the principles of quantum mechanics ascertain the detailed structure and behavior of atoms and molecules. By employing wavefunctions and the Schrödinger equation, quantum chemistry enumerates the energy levels, electronic configurations, and spectral properties fundamental to molecular systems. Through electronic structure theory, quantum chemistry elucidates molecular interactions and reactivity pathways that underlie chemical bonding and molecular properties. This branch of physical chemistry deeply informs models of molecular dynamics by accounting for variables such as electron spin, orbital hybridization, and resonance structures.

Statistical mechanics constitutes another cornerstone of physical chemistry by providing the link between microscopic properties of molecules and their macroscopic manifestations. This essential domain considers the aggregate behavior of large ensembles of molecules, deriving statistical averages to explain thermodynamic properties. Through Boltzmann statistics and partition functions, statistical mechanics elucidates temperature dependence, heat capacities, and phase transitions. It further predicates molecular motion and distribution laws, such as those depicted by the Maxwell-Boltzmann distribution,

1.1. DEFINITION AND SCOPE OF PHYSICAL CHEMISTRY

closely connected with kinetic theory and chemical kinetics.

Chemical kinetics, within the scope of physical chemistry, examines the rates of chemical reactions and the factors influencing these rates. It encompasses the study of reaction mechanisms by establishing rate laws and equations derived from empirical and theoretical analysis. The Arrhenius equation, transition state theory, and collision theory provide a theoretical framework for elucidating activation energies and identifying reaction intermediates. This aspect of physical chemistry is crucial for understanding catalytic processes, reaction networks, and the interplay of temperature and concentration on reaction velocities.

Physical chemistry also engages directly with spectroscopy and photochemistry to reveal critical insights into molecular structures and reaction dynamics. Spectroscopic techniques, including infrared, UV-Visible, NMR, and mass spectrometry, present powerful tools for probing the interactions of electromagnetic radiation with matter. These techniques afford precise determination of molecular geometries and conformational analysis, enabling the deconstruction of complex chemical signatures. In parallel, photochemistry assesses the influence of light on chemical systems, unveiling reaction pathways activated by photon absorption and emission processes.

The scope of physical chemistry is inherently interdisciplinary, bridging fields such as materials science, environmental chemistry, and biochemical systems. The fundamental principles elucidated by physical chemistry underpin the understanding and innovation in areas such as nanotechnology, catalysis, and pharmaceutical development. Its role extends to exploring the physical properties of new materials, including polymers, alloys, and composites, where phase diagrams and thermodynamic stability are of paramount importance.

In environmental science, physical chemistry plays a decisive role in understanding transport phenomena, atmospheric chemistry, and pollutant interactions. Mechanistic insights gleaned from physical chemistry drive models predicting the fate and behavior of chemical species within ecosystems, influencing strategies for pollution control and remediation. The balance and the interplay of chemical equilibria and kinetics are pivotal in analyzing solubility patterns and reaction equilibria in natural waters and soil matrices.

In biophysical chemistry, the scope extends to elucidating the intricate structures and functions of biological macromolecules—proteins, nucleic acids, lipids, and carbohydrates. The application of thermodynamic principles and kinetic models helps in dissecting enzyme action, molecular recognition, and signal transduction mechanisms. Quantum chemical methods provide essential insights into protein folding, conformational changes, and the electronic character of active sites within biomolecules.

Physical chemistry is also pivotal in the burgeoning field of computational chemistry, using sophisticated algorithms to simulate and predict the properties of molecules and materials. Through molecular dynamics and Monte Carlo simulations, physical chemistry explores energy landscapes, molecular interactions, and reaction pathways with a high degree of fidelity. The integration of computational models with experimental data facilitates the accelerated discovery and design of molecular structures with desired physical and chemical properties.

The modern landscape of physical chemistry continues to expand with technological advancements, where high-resolution imaging techniques and analytical instrumentation enhance the ability to observe and manipulate chemical systems at unprecedented scales. Innovations such as atomic force microscopy, X-ray crystallography, and electron microscopy enable a deeper exploration of material surfaces and nanostructures, enriching the physical chemist's toolkit for structural analysis.

At its core, physical chemistry remains a vibrant and dynamic field, defined by its rigorous application of physical theories to understand the subtleties of chemical systems. It persistently refines the theoretical approaches and experimental techniques that bridge particles and systems, clarifying the interactions governing chemical behavior. Physical chemistry provides a lens through which one can systematically analyze the principles underlying chemical phenomena, underscoring the universality of chemical change in the natural world. Its vast scope continues to push the boundaries of scientific enquiry and technological progression, firmly establishing its role as an essential pillar of modern chemistry.

1.2 Historical Development and Theories

The field of physical chemistry has evolved through the convergence of pioneering ideas, empirical discoveries, and the construction of theoretical frameworks. Its historical development is a testament to the ingenuity and perseverance of scientists who have sought to understand the fundamental principles governing chemical systems. Tracing the chronological arc of its evolution offers insights into the key milestones and scientific revolutions that have shaped

1.2. HISTORICAL DEVELOPMENT AND THEORIES

its current landscape.

The roots of physical chemistry can be traced back to the late 18th and early 19th centuries, a period marked by transformative advancements in chemistry and physics. Antoine Lavoisier's seminal work in establishing the law of conservation of mass laid the groundwork for systematic chemical analysis, dispelling the phlogiston theory and fostering a more quantitative approach to chemistry. This era marks the beginning of the search for a deeper understanding of chemical processes, one that aligns with core physical principles.

Simultaneously, the development of atomic theory gained traction through John Dalton's assertions in the early 19th century, positing that elements comprise indivisible atoms with distinct weights. Dalton's atomic theory set the stage for subsequent inquiries into atomic interactions and provided a structural foundation for interpreting chemical phenomena. These early advancements underscored the need for a scientific approach that bridged atomic-level understanding with macroscopic observations.

In the mid-19th century, physical chemistry emerged more distinctly as a separate discipline through the convergence of thermodynamics and kinetic theory. Rudolf Clausius and William Thomson (Kelvin) pioneered the foundational principles of thermodynamics, establishing a quantitative understanding of energy transformation in chemical systems. Clausius introduced the concept of entropy, a revolutionary idea that formalized the directionality of processes and linked energy dispersal to the second law of thermodynamics. This conceptualization was pivotal in explaining why certain chemical processes are spontaneous, leading to a deeper comprehension of equilibrium states.

The kinetic theory of gases, elucidated by figures such as James Clerk Maxwell and Ludwig Boltzmann, further advanced the understanding of molecular motion and distributions. Maxwell's development of the speed distribution for gas molecules demonstrated the link between molecular velocities and macroscopic properties, effectively merging the micro-macro dichotomy. Boltzmann's statistical interpretation of entropy reinforced the probabilistic nature of thermodynamic laws, laying the groundwork for statistical mechanics. This framework illuminated how energy partitioning among microscopic constituents manifested in observable macroscopic properties, thus revolutionizing analytical approaches in physical chemistry.

The advent of quantum mechanics in the early 20th century marked a paradigm shift, refining the theoretical understanding of chemical bonding and spectroscopy within physical chemistry. Max Planck's introduction of quantized energy levels challenged classical continuous models and inaugurated the quantum era. Albert Einstein's explanation of the photoelectric effect, and Niels Bohr's model of the hydrogen atom, further solidified the quantum perspective, elucidating the discrete nature of electronic transitions and providing insights into atomic spectra.

Erwin Schrödinger's wave mechanics and Werner Heisenberg's matrix mechanics provided complementary formulations for describing electron behavior within atoms and molecules. Schrödinger's wave equation enabled the calculation of molecular orbitals, fostering the development of quantum chemistry as a sub-discipline within physical chemistry. This theoretical apparatus allowed chemists to predict and explain the electronic structure of complex molecules, thus influencing analyses of reactivity and chemical kinetics.

The consolidation of wave-particle duality by Louis de Broglie and the uncertainty principle by Heisenberg further enriched quantum theory, cementing its centrality to physical chemistry. These theories addressed the limitations of classical mechanics in representing the behavior of subatomic particles, providing a coherent framework for exploring the intrinsic probabilistic nature of particles.

Quantum chemistry's emergence also prompted the development of advanced computational methodologies, transforming physical chemistry into a discipline where theoretical and computational insights drive experimental inquiries. The implementation of Hartree-Fock and post-Hartree-Fock methods in the mid-20th century facilitated the computational prediction of molecular properties and reactions, thereby integrating computational power with traditional chemical analysis.

Parallel to quantum theory, the historical trajectory of physical chemistry has been marked by the expanding applications of spectroscopy and the refinement of spectroscopic techniques. With roots as far back as the 19th century, spectroscopic methods have elucidated molecular vibrations, rotations, and electronic arrangements, providing direct evidence for quantum mechanical descriptions. Infrared (IR) and nuclear magnetic resonance (NMR) spectroscopy, among others, have become indispensable tools in the physical chemist's arsenal, allowing for precise structural determinations and dynamic studies.

The refinement of these techniques has gone hand in hand with the formulation of new theories. For example, Linus Pauling's work on chemical bonding and hybridization theory synthesized quantum-mechanical principles with empiri-

1.2. HISTORICAL DEVELOPMENT AND THEORIES

cal observations, offering a framework to understand molecular geometry and hybrid orbitals. His resonance theory further advanced concepts of electron delocalization, enhancing the comprehension of complex conjugated systems and reaction mechanisms.

Physical chemistry's evolution also reflects its integration with interdisciplinary fields, extending its impact beyond traditional chemical systems. Notably, the principles gleaned from advancing physical chemistry have influenced the development of catalytic processes and materials science. The work of Fritz Haber and Carl Bosch on the ammonia synthesis through heterogeneous catalysis exemplifies how thermodynamic and kinetic insights can transform industrial chemistry, underscoring physical chemistry's practical importance.

As the 20th century progressed, physical chemistry continued to diversify through new theoretical models, computational techniques, and experimental methodologies. The growth of spectroscopy and microscopy, alongside advancements in crystallography and surface analysis, has constantly expanded the ability to probe chemical systems at elemental and molecular levels. This growth reflects in the understanding of complex systems, such as biomolecular interactions and quantum dots, offering profound implications for biochemistry and nanotechnology.

Moreover, the rise of non-equilibrium thermodynamics and chaos theory has furthered the exploration of dynamic systems far from equilibrium, broadening the applicability of physical chemistry to biological and environmental systems. These theories accentuate the unpredictability and intricacies associated with dynamic transformations and their stabilizing energy exchanges, pushing the boundaries of conventional thermodynamic understanding.

The trajectory of physical chemistry is a testimony to the powerful interplay between theoretical breakthroughs and empirical discoveries, an ongoing endeavor to refine and expand the conceptual and methodological foundations of chemical analysis. The historical development of physical chemistry reflects an evolving discipline characterized by the cumulative contributions of scientists across generations, steadily unlocking the mysteries of chemical behavior and interactions. The rich legacy of historical theories and innovations not only explains past phenomena but also provides the scaffolding upon which future inquiries into chemical systems may flourish, continuing to propel the field into new intellectual territories.

1.3 Mathematical Foundations

The mathematical foundations of physical chemistry form an intricate lattice upon which theoretical models and empirical observations are constructed. These mathematical tools provide the language and framework necessary for the quantitative analysis of chemical phenomena, enabling precise descriptions and predictions of chemical behavior. This section delves into the essential mathematical concepts and techniques indispensable to physical chemistry, highlighting their application to chemical systems.

At the core of physical chemistry's mathematical framework lies calculus, both differential and integral, which serves as a critical tool for describing changes within chemical systems. Differential calculus provides the means to express and analyze rates of change, which is fundamental in understanding reaction kinetics and dynamics. The rate laws derived from empirical data express rate as a differential equation, such as:

$$\frac{d[A]}{dt} = -k[A]^n$$

where $[A]$ is the concentration of reactant A, k is the rate constant, and n indicates the reaction order. Solving these equations offers insights into how concentration changes with time, unraveling the kinetics of reactions.

Integral calculus, conversely, plays a vital role in quantifying accumulated changes, such as total reaction extent or areas under curves in spectroscopic data. For example, integrating a rate law provides the concentration of species as a function of time, expressed as:

$$[A] = [A]_0 e^{-kt}$$

In equilibrium thermodynamics, calculus is utilized to derive thermodynamic properties from fundamental equations of state, bridging changes at minor and infinitesimal scales with macroscopic observables. Partial derivatives, a cornerstone of multivariable calculus, are extensively employed in these derivations. The partial derivatives help define properties such as volume, pressure, and the Gibbs free energy across different conditions, with an emphasis on concepts such as the Maxwell relations and the Jacobian.

Linear algebra is another foundation of physical chemistry, providing avenues to explore molecular orbital theory, spectroscopy, and quantum states. Linear

1.3. MATHEMATICAL FOUNDATIONS

algebra enables the description of quantum states as vectors in a multidimensional space, with matrices representing operators on these states. Hamiltonian operators encapsulate total energy, with eigenvalues and eigenvectors offering insight into quantized energy levels and wavefunctions:

$$\hat{H}\psi = E\psi$$

where $\hat{H}$ denotes the Hamiltonian, E is the energy eigenvalue, and ψ represents the wavefunction. The diagonalization of matrices, finding eigenvalues and eigenvectors, forms the backbone of molecular structure prediction and analysis within computational chemistry.

Furthermore, group theory, an abstract algebra branch, plays a pivotal role in symmetry analysis within molecules. Employing group theory, chemists systematically classify molecular vibrations, predict spectroscopic transitions, and understand symmetry-forbidden processes, enhancing their insights into molecular spectra and reactivity patterns.

The use of statistics and probability forms an integral part of physical chemistry, mainly through statistical mechanics, which connects microscopic behavior to macroscopic properties. Probability distributions, such as the Boltzmann distribution, model the occupancy of particles across energy states:

$$P_i = \frac{e^{-\beta E_i}}{Z}$$

where P_i represents the probability of a particle being in state i with energy E_i, $\beta = 1/k_B T$ is the inverse temperature (with k_B being the Boltzmann constant), and Z denotes the partition function, summing over all states. Boltzmann statistics, Fermi-Dirac, and Bose-Einstein distributions extend these ideas across diverse quantum statistical ensembles, relevant to distinct categories of particles.

Numerical methods are also essential, particularly in scenarios where analytical solutions are unattainable. Techniques such as finite difference methods, Monte Carlo simulations, and molecular dynamics provide approximate solutions to complex systems, allowing molecular properties to be deduced under real-world conditions. Computationally, algorithms iterate over vast datasets to refine energy landscapes, simulate diffusion processes, and solve multidimensional integrals.

In electrochemistry, differential equations and complex analysis are harnessed to study ionic conductance, transport phenomena, and interface potential differences. The Nernst equation, derived via thermodynamic and electrochemical principles, exemplifies the intersection of mathematics with chemical electrostatic potential:

$$E = E^\circ - \frac{RT}{nF} \ln \frac{[red]}{[ox]}$$

where E represents electromotive force, E° is standard potential, R is the gas constant, T temperature, n moles of electrons exchanged, F is the Faraday constant, and $[red]$ and $[ox]$ are concentrations of reduced and oxidized species.

Fourier analysis emerges as indispensable in processing spectroscopic data, as it transforms time-domain signals into frequency-domain spectra. This analysis is crucial in disentangling complex overlapping signals, instrumental in techniques such as NMR, IR, and mass spectroscopy.

Vector calculus and tensor algebra play significant roles when addressing dynamic systems and spatial diffusion, epitomized by vector fields and flow gradients relating to flux, including current density and magnetic field interactions. Tensor calculus extends these ideas into more intricate deformation and anisotropic systems, regularly applied within material physics and crystallography.

The conversion between different unit systems and dimensional analysis is another keystone within physical chemistry's mathematical practice, enabling the coherent, consistent application of equations across varied scales and disciplines. These methods assess validity, deduce scaling laws, and evaluate feasibility, providing clarity and predictability in experimental and theoretical approaches.

The complex interplay and application of these mathematical principles lay the groundwork for groundbreaking advancements in physical chemistry, facilitating the exploration of new chemical spaces. By enabling the concise description of diverse phenomena—from single-molecule electron states and energy transitions to macroscopic thermodynamic anomalies—mathematics remains intrinsic to the continuous unveiling of chemical science's mysteries.

As we look toward future developments, the integration of artificial intelligence and machine learning within physical chemistry increasingly relies on mathematical rigor to unravel complex data structures and predict novel chemical reactions. Through innovative algorithmic solutions, such as pattern recognition and neural networks, the field is witnessing the synthesis of traditional mathematical principles with cutting-edge computational technologies to extend its predictive capabilities and revolutionary findings.

In sum, the mathematical foundations underpinning physical chemistry are robust and multifaceted, offering indispensable tools for the exploration and understanding of the structure, behavior, and interactions of chemical systems. These foundations not only support the current landscape of research and application but also fortify the field's progress in unraveling future chemical phenomena, ensuring the continuous evolution and expansion of knowledge at the intersection of mathematics and chemistry.

1.4 Units, Measurements, and Error Analysis

In physical chemistry, the precision and accuracy of data are vital to testing hypotheses, verifying models, and advancing scientific knowledge. Therefore, a thorough understanding of units, measurements, and error analysis is indispensable. This section delves into the standardized units in chemistry, methods of precise measurement, and techniques for error analysis, emphasizing their critical functions in sustaining the integrity and reproducibility of scientific inquiry.

The International System of Units (SI) forms the bedrock of scientific measurement, providing a standardized framework for expressing fundamental quantities. In physical chemistry, key SI units are employed for mass (kilogram, kg), length (meter, m), time (second, s), amount of substance (mole, mol), temperature (kelvin, K), electric current (ampere, A), and luminous intensity (candela, cd). These units provide consistency and comparability across experiments and disciplines.

Beyond the basic units, derived units are extensively utilized in physical chemistry to express quantities such as energy (joule, J), pressure (pascal, Pa), and volume (liter, L, though strictly not SI), facilitating communication and calculations. For instance, 1 joule is defined as 1 kilogram meter squared per second squared ($1 \text{ J} = 1 \text{ kg m}^2 \text{ s}^{-2}$), encapsulating the fundamental relationship between work and energy.

Conversions between units are routine in chemical analyses, necessitating proficiency with conversion factors. For example, the conversion between temperature scales, such as Celsius to Kelvin, is straightforward due to the linear relationship:

$$T(K) = T(°C) + 273.15$$

In more complex conversions, dimensional analysis proves invaluable, ensuring that calculations retain dimensional consistency, thereby reducing the risk of errors in interconverting units.

The precision of measurements hinges on the accuracy and sensitivity of the instruments employed. Instruments cater to different types of measurements, ranging from simple hand-held devices like pipettes to sophisticated equipment such as spectrometers and calorimeters. Precision is determined by repeatability, while accuracy involves how closely a measured value aligns with a true value. Calibration against known standards ensures that instruments provide accurate and reliable data. For instance, the calibration of spectrophotometers using standard absorbance solutions aids in determining concentrations via Beer-Lambert law:

$$A = \varepsilon c l$$

where A is absorbance, ε the molar absorptivity, c the concentration, and l the path length.

Error analysis is integral to interpreting physical and chemical measurements. It encompasses identifying error sources, estimating error magnitude, and minimizing their impact. Errors typically divide into three types: systematic, random, and gross.

Systematic errors arise from identifiable sources, leading to consistent deviation from the true value. Calibration mishaps, instrumental bias, and methodological flaws are primary causes. Addressing systematic errors involves comprehensive instrument calibration, methodological refinement, and adopting reference materials to ensure traceability. An example includes using certified reference materials for pH meter calibration, reducing bias in pH-dependent analyses.

Random errors stem from unpredictable variations affecting measurement precision. These errors follow a probability distribution, often modeled by normal distributions, where standard deviation (σ) illustrates the spread of data about the mean ($\bar{x}$):

$$\sigma = \sqrt{\frac{\sum (x_i - \bar{x})^2}{N - 1}}$$

1.4. UNITS, MEASUREMENTS, AND ERROR ANALYSIS

Mitigating random errors requires repetition, producing a statistical increase in precision by averaging measurements.

Gross errors, occasionally termed blunders, result from operator mistakes such as transcription errors or misreading scales. These are preventable through operator training, meticulous procedure adherence, and quality control processes.

The propagation of errors, when manipulating measurements through calculations, requires detailed attention. The method involves combining individual measurement errors to estimate overall uncertainty in derived results. For addition or subtraction, errors combine linearly:

$$\Delta Q = \sqrt{(\Delta A)^2 + (\Delta B)^2}$$

For multiplication or division, relative errors combine:

$$\frac{\Delta Q}{Q} = \sqrt{\left(\frac{\Delta A}{A}\right)^2 + \left(\frac{\Delta B}{B}\right)^2}$$

where Q represents the calculated quantity, A and B are measurements, and Δ denotes their errors.

Significant figures play a significant role in expressing measurement precision, reflecting the certainty in the reported digits. They communicate error magnitude and reduce misrepresentation of precision. Simplified rules ensure uniformity in displaying significant figures in calculations, such as rounding off at the least number of significant figures across multiplied quantities.

Graphical analysis provides visual insights into error extents, reproducibility, and underlying trends. Error bars, standard deviation, and confidence intervals illustrate data variability and reliability. For example, linear regression analysis acknowledges potential errors by providing standard errors and R^2 values to determine fit quality and predict reliable trends.

Data interpretation demands rigorous statistical analysis. Hypothesis testing, correlation analysis, and ANOVA (Analysis of Variance) authenticate relationships between variables and determine significance. In experimental design, factorial designs and response surface methodologies allow comprehensive, systematic exploration of parameter spaces, enhancing reproducibility.

Control charts, a quality assurance staple, monitor experimental processes over time by charting points within control limits. These charts provide early identification of trends and aberrations, ensuring experiments remain stable and free from undue variability. This method is essential in kinetic monitoring, where deviations might signify changes in reaction conditions or substrate availability.

The integration of sophisticated software in data analysis, such as chemometrics and data mining, enriches error analysis further. Tools like principal component analysis (PCA) and partial least squares (PLS) refine data sets by reducing dimensionality and identifying core patterns amidst noise, which proves invaluable in complex systems such as bioinformatics and spectroscopic data evaluation.

In sum, units, measurements, and error analysis constitute the fundamental framework securing data's veracity and credibility in physical chemistry. By implementing standardized units, applying rigorous measurement techniques, and executing comprehensive error analysis, the discipline sustains both the precision and accuracy demanded by contemporary chemical inquiry. This methodological rigor continues to support advances in understanding chemical systems, ensuring findings withstand scrutiny and offering replicability across diverse scientific endeavors. As the field progresses, integrating new technologies and methodologies will further refine these practices, perpetuating the quest for deeper insights and novel discoveries.

1.5 Interdisciplinary Connections

The realm of physical chemistry is intrinsically interdisciplinary, interweaving its principles with a myriad of scientific fields such as physics, biology, engineering, environmental science, and materials science. This interplay not only broadens the horizons of physical chemistry but also enriches the collaborative potential among disciplines, fostering innovation and discovery. This section explores the profound connections of physical chemistry with these varied fields, elucidating their synergistic relationships and their impact on advancing scientific understanding and technological progress.

The connection between physical chemistry and physics is foundational, as physical chemistry initially emerged from the application of physical principles to chemical problems. This intersection is prominently manifested in quantum chemistry and thermodynamics. Quantum mechanics, a cornerstone of both physical chemistry and physics, provides a theoretical framework for understanding the electronic structure of atoms and molecules. It facili-

1.5. INTERDISCIPLINARY CONNECTIONS

tates the calculation of molecular orbitals, energy levels, and transition states, which are crucial for predicting chemical reactivity and bonding patterns.

The principles of thermodynamics are equally shared between the two disciplines. Physical chemistry utilizes thermodynamic laws to explain energy transformations and phase equilibria within chemical systems. Concepts such as entropy, enthalpy, and free energy are pivotal in predicting reaction spontaneity and directionality, properties that reflect the inherent nature of physical systems. Collaborative research often focuses on unraveling novel energy conversion and storage mechanisms, such as those in fuel cells and batteries, where thermodynamic efficiency is paramount.

In the domain of photochemistry, physicists and physical chemists collaborate to explore the interactions between light and matter, a crucial aspect for developing optical materials and devices. Techniques like laser spectroscopy arise from these collaborations, providing tools to probe and manipulate particles at atomic and molecular levels, impacting fields as diverse as telecommunications and quantum computing.

The interactions between physical chemistry and biology have given birth to biophysical chemistry, a field that applies physical chemical principles to elucidate biological processes and mechanisms. Physical chemists investigate the structural dynamics and functions of biomolecules, such as proteins and nucleic acids, employing techniques like X-ray crystallography and NMR spectroscopy. These studies contribute to understanding molecular folding, binding interactions, and enzyme catalysis.

Thermodynamics and kinetics play significant roles in biochemistry, explaining processes such as enzyme-substrate interactions and metabolic pathways. The Michaelis-Menten model, a quantitative formulation originating from kinetic theory, provides insight into enzyme activity and its regulation by competitive and non-competitive inhibitors. This understanding underpins drug design and the development of therapeutic agents targeting biochemical pathways in diseases.

Nanotechnology and materials science represent burgeoning areas where the fusion of physical chemistry principles with engineering and physics has catalyzed remarkable advancements. At the nanoscale, the quantum mechanical properties of materials diverge significantly from their bulk characteristics, demanding a thorough understanding of quantum states and electron configurations as explored in physical chemistry.

The insights derived from physical chemistry into reaction kinetics and surface interactions are crucial for the development of catalysts and nanomaterials. Techniques such as atomic layer deposition and self-assembly leverage this understanding to fabricate materials with specific electronic, optical, and mechanical properties, instrumental in semiconductor technology and the development of medical diagnostics.

In environmental science, physical chemistry contributes substantially to understanding the chemical behavior of ecosystems and pollutants. The study of atmospheric chemistry, for instance, relies on physical chemistry principles to model the reactions and transport processes of chemical species in the atmosphere. Photochemical smog formation, greenhouse gas interactions, and ozone layer depletion are phenomena extensively researched using kinetic models and spectroscopic techniques.

Surface chemistry, a subfield of physical chemistry, is pivotal in studies of soil and water chemistry. It informs the understanding of adsorption processes, pollutant interactions, and nutrient cycles in environmental systems. These insights guide the development of remediation strategies and technologies, such as catalysis for environmental cleanup and design of sustainable materials.

Chemical engineering and physical chemistry are closely linked by their mutual focus on transforming raw materials into valuable products through chemical reactions. Chemical engineers utilize the principles of thermodynamics, kinetics, and transport processes, all grounded in physical chemistry, to design and optimize industrial processes. The development of chemical reactors, separation techniques such as distillation and chromatography, and process simulations are rooted in physical chemical methodologies.

The cross-disciplinary partnerships also extend to the medical and pharmaceutical fields, where physical chemistry's understanding of molecular interactions and thermodynamics aids in drug formulation and delivery mechanisms. Liposomal drug carriers, for example, capitalize on bilayer thermodynamics and surface chemistry principles to enhance drug stability and delivery efficacy.

Moreover, advancements in computational physical chemistry have empowered predictive modeling across disciplines, enhancing the capability to simulate complex biological systems, materials properties, and chemical processes. These computational techniques enable the exploration of molecular dynamics and reactivity landscapes, offering predictions that guide experimental design and innovation.

The intersection of physical chemistry with data science is increasingly significant, as vast datasets from simulations and experiments require sophisticated analytical tools. Techniques such as machine learning and artificial intelligence are being integrated to identify patterns, optimize reactions, and predict material properties, heralding a new era of data-driven discovery in physical chemistry.

- The interdisciplinary connections of physical chemistry extend its impact across the scientific spectrum, fostering a collaborative environment conducive to innovation and discovery.
- These interactions not only enhance our understanding of fundamental chemical phenomena but also translate into practical applications that address global challenges in energy, health, and sustainability.
- The reciprocal flow of knowledge and techniques between physical chemistry and other disciplines exemplifies the dynamic and integrative nature of modern science, paving the way for future advancements and technological breakthroughs.

1.6 Applications and Importance

Physical chemistry stands as a fundamental pillar in scientific research and industrial applications, offering profound insights into the molecular mechanisms underpinning a multitude of chemical processes. The discipline's significance is not confined to theoretical exploration; it manifests across diverse practical applications, catalyzing advancements in technology, medicine, environmental sustainability, and industrial processes. This section delves into the extensive applicability and critical importance of physical chemistry, highlighting its transformative impact across various sectors.

The pharmaceutical industry exemplifies the robust applications of physical chemistry, where understanding molecular interactions is paramount for drug design and development. Techniques such as computational chemistry and molecular modeling, grounded in quantum chemistry and thermodynamics, allow researchers to predict pharmacodynamic interactions and optimize drug candidates. Physical chemistry aids in elucidating the structures and energetics of drug-receptor interactions, a crucial factor in determining efficacy and specificity. Quantum mechanics often contributes to structure-activity relationships, providing a theoretical basis for the rational design of therapeutics targeting specific enzymes or receptors.

Moreover, the principles of chemical kinetics and thermodynamics are integral to drug formulation, impacting solubility, stability, and release rates. The development of controlled-release mechanisms leverages kinetic models to regulate active ingredient delivery over extended periods, enhancing therapeutic outcomes while minimizing side effects. For instance, polymer-based drug delivery systems make use of diffusion and degradation kinetics, as described by the Higuchi and Korsmeyer-Peppas models, to tailor release profiles.

In materials science, the insights from physical chemistry fuel the synthesis and characterization of novel materials with unique properties. Understanding the energetics and kinetics of crystallization processes facilitates the design of advanced materials with applications in electronics, catalysis, and energy storage. Physical chemistry is instrumental in manipulating bandgaps, surface properties, and structural phases, thereby influencing material characteristics such as conductivity, magnetism, and optical behavior.

Nanotechnology, with its potential to revolutionize material fabrication, relies heavily on the principles of physical chemistry. Quantum dots, nanowires, and graphene are engineered by applying quantum mechanical models to control electron behavior and surface chemistry at the nanoscale. These nanomaterials, characterized using spectroscopic and microscopic techniques derived from physical chemistry, find applications in solar cells, sensors, and medical diagnostics, highlighting the interdisciplinary impact of the field.

In catalysis, physical chemistry enables the development of catalysts that accelerate chemical reactions, crucial for industrial processes ranging from petrochemicals to polymers. By elucidating surface interactions and reaction pathways, physical chemistry guides the optimization of catalysts, enhancing selectivity and reducing activation energies. Techniques such as temperature-programmed desorption and X-ray photoelectron spectroscopy provide insights into catalyst structure and function, informing the development of more efficient catalytic systems.

Environmental science witnesses the application of physical chemistry through the study of atmospheric and aquatic chemistry, where reaction kinetics and thermodynamics are vital for understanding pollutant behavior and transformation. The modeling of atmospheric reactions, such as photochemical smog formation and ozone depletion, relies on kinetics to predict the fate of chemical species and assess environmental impact.

1.6. APPLICATIONS AND IMPORTANCE

Physical chemistry principles underlie the design of scrubbers and adsorbents for pollutant removal, contributing to air and water purification technologies crucial for environmental sustainability.

Energy conversion and storage technologies, essential for addressing global energy demands, are deeply rooted in the insights provided by physical chemistry. Fuel cells, batteries, and photovoltaic devices benefit from the understanding of electron transfer processes, ion diffusion, and interfacial phenomena. In batteries, the kinetics of electrode reactions and diffusion across membranes determine efficiency and capacity, with research focusing on materials that enhance these parameters.

The development of renewable energy technologies, such as solar cells, employs physical chemistry to improve light absorption, charge separation, and electron transport. Photophysics and photoelectrochemistry, branches influenced heavily by physical chemistry, contribute to optimizing these processes, striving for higher conversion efficiencies and cost-effective production.

In academia and research, advances in physical chemistry continue to drive forward our fundamental understanding of chemical phenomena. Sophisticated experimental techniques, such as ultrafast spectroscopy and single-molecule electronics, are rooted in physical chemistry and allow for the study of transient states and molecular dynamics with unprecedented detail. These methodologies provide insights into reaction mechanisms and energy transfer processes, underpinning discoveries ranging from reactive intermediates to coherent control.

The field's mathematical and theoretical frameworks, encompassing quantum mechanics and statistical thermodynamics, guide the interpretation of experimental findings and the prediction of chemical behavior. This synergy between theory and experiment strengthens the predictive power of physical chemistry, fostering innovation across scientific domains.

The scope of physical chemistry extends to the field of agriculture, where it informs the development of agrochemicals and fertilizers. Understanding the kinetics and reaction pathways of active ingredients aids in optimizing formulations to enhance efficacy while minimizing environmental impact. Soil chemistry, where ion exchange and adsorption processes are studied, benefits from insights into thermodynamic equilibria, influencing agricultural productivity and sustainability.

In sum, the applications and importance of physical chemistry are vast and varied, underscoring its role as a critical enabler of scientific and technological advancements. By bridging molecular-level understanding with macroscopic observables, physical chemistry offers a profound comprehension of natural phenomena and engineered systems, impacting industries and research fields globally. Its ongoing evolution and integration with emerging technologies ensure that physical chemistry remains at the forefront of addressing contemporary challenges, from sustainable energy solutions and environmental protection to advancing healthcare and material development. As the field continues to advance, the potential for new applications expands, setting the stage for further breakthroughs that will shape the future of science and technology.

Chapter 2

Quantum Chemistry and Atomic Structure

Quantum chemistry provides the essential framework for understanding the electronic structure and behavior of atoms and molecules, probing the fundamental nature of matter through principles rooted in quantum mechanics. This chapter delves into the Schrödinger equation and explores the concept of atomic orbitals, offering insight into how quantum numbers describe atom configurations. By examining approximation methods and their application in molecular systems, readers will grasp the theoretical constructs that predict and explain chemical bonding patterns and reactivity. Through the lens of spectroscopy, quantum chemistry connects foundational theory to experimental practice, facilitating the demystification of atomic and molecular spectra, which unveils the intrinsic properties of substances.

2.1 Fundamentals of Quantum Mechanics

Quantum mechanics represents a significant departure from classical physics, offering a precise mathematical framework to describe the smallest particles of matter and energy. Central to its principles are concepts like wave-particle duality and the uncertainty principle, which challenge our classical intuitions about nature.

Wave-particle duality suggests that every particle or quantum entity may be described as either a particle or a wave. This duality is beautifully captured by experiments such as the double-slit experiment, where particles like electrons produce an interference pattern, a behavior characteristic of waves, when not observed, yet exhibit particle-like properties in the presence of an observer. The mathematics that support this duality stems from the de Broglie hypothesis, which postulates that every particle with momentum has an associated wavelength, $\lambda = \frac{h}{p}$, where h is Planck's constant and p is the momentum. This foundational principle is essential for understanding complex phenomena in quantum mechanics, including the nature of atomic and molecular structures.

Heisenberg's uncertainty principle further adds to the wave-particle duality by imposing fundamental limits on what can be known about a particle's properties. Specifically, it states that the more precisely a particle's position is known, the less precisely its momentum can be known, and vice versa. Mathematically, this is expressed as:

$$\Delta x \Delta p \geq \frac{\hbar}{2}$$

where Δx and Δp represent the uncertainties in position and momentum, respectively, and $\hbar$ is the reduced Planck's constant. This uncertainty is not due to the limitations of measuring instruments but is an inherent aspect of quantum systems, suggesting a completer understanding of phenomena at scales much smaller than those we deal with in everyday life.

The mathematical framework of quantum mechanics encompasses linear algebra and differential equations, most notably through the formalism of wave functions. The wave function, usually denoted $\psi(x, t)$, is a fundamental concept that provides a description of the quantum state of a system. It is complex-valued and contains all the probabilistic information about a particle's position and momentum. The probability density for finding a particle at position x at time t is given by $|\psi(x, t)|^2$.

An essential development in quantum mechanics is the concept of operators, which correspond to observable physical quantities. For instance, the position operator $\hat{x}$ acts on the wave function to yield position information, while the momentum operator is given by:

2.1. FUNDAMENTALS OF QUANTUM MECHANICS

$$\hat{p} = -i\hbar \frac{\partial}{\partial x}$$

and acts on the wave function to find momentum information. These operators are crucial in the formulation of the quantum mechanical Hamiltonian, which embodies the total energy of the system, akin to mechanics in classical physics, but here it acts as a differential operator.

The superposition principle is another pillar of quantum mechanics, stating that if ψ_1 and ψ_2 are two possible states of a system, the linear combination $a\psi_1 + b\psi_2$, where a and b are complex numbers, is also a valid state. This principle explains phenomena such as quantum interference and is foundational for understanding quantum entanglement, where particles become correlated in ways not explainable by classical means alone.

Quantum entanglement, a phenomenon where quantum states of individual particles become interdependent regardless of the distance separating them, underpins much of the discussion about the non-locality of quantum mechanics. When particles are entangled, the state of one (no matter how distant) instantaneously affects the state of another, as demonstrated in experiments testing Bell's inequalities. These experiments confirm that quantum mechanics cannot be merely a local, deterministic theory, challenging the classic Einstein-Podolsky-Rosen (EPR) argument for the completeness of quantum mechanics.

Yet another fascinating outcome of quantum mechanics is the principle of quantization itself, which imposes discrete values, or "quanta," for physical quantities like energy. This discrete nature arises naturally when solving the Schrödinger equation (a topic further explored in subsequent sections for various systems). For instance, the quantization of energy levels in an atom leads directly to its stability and the spectral lines observed in atomic spectra.

These fundamental principles converge in the concept of quantum tunneling, an inherently quantum phenomenon where particles have a non-zero probability of crossing potential barriers even when they lack the classical energy to do so. Tunneling has profound implications in fields ranging from nuclear fusion to the operation of modern electronic devices, such as semiconductors and tunnel diodes.

Illustrative of this extensively comprehensive field is the quantization observed in a particle confined within a potential box, a problem known as the quantum mechanical particle in a box or infinite potential well. In this model, the wave function must vanish at the boundaries, suggesting only a discrete set of wave functions and associated energies – a clear depiction of quantized phenomena.

The evolution of quantum systems is governed by the Schrödinger equation, an indispensable tool in quantum mechanics, used for predicting how a quantum state changes over time. This differential equation emerges in different formulations, including the time-dependent and time-independent versions, with the latter crucial for stationary states that form the basis for much of quantum chemistry and atomic physics. The time-independent Schrödinger equation is represented as:

$$\hat{H}\psi = E\psi$$

where $\hat{H}$ is the Hamiltonian operator, ψ is the wave function of the system, and E represents the energy eigenvalue associated with the state. The solutions to this equation guide us to quantized energy levels and intricate wave functions that characterize a system's state, defining our understanding of atomic and molecular structures.

Despite the exotic nature of its tenets, quantum mechanics has shown remarkable congruence with reality, offering precise predictions verified repeatedly in experiments. Its applications range widely, including our understanding of chemical bonding, the electronic behavior of materials, advances in quantum computing, and even the cosmological fabric of the universe.

The principles of quantum mechanics underscore a nuanced and sophisticated view of the physical universe, one in which probability and uncertainty are inherent, duality pervades, and interconnections transcend familiarity. They provoke profound philosophical questions while providing insight into the natural world, establishing a foundation upon which modern scientific inquiry continues to flourish and expand.

2.2 The Schrödinger Equation

The Schrödinger equation stands as a cornerstone of quantum mechanics, providing the mathematical framework essential for describing the evolution and behavior of quantum systems. It bridges the wave mechanics perspective with quantized physical observations, offering a complete picture of the probabilistic nature of subatomic particles.

2.2. THE SCHRÖDINGER EQUATION

Central to this is the concept of the wave function, ψ, which encapsulates all the information necessary to describe a quantum system. The Schrödinger equation, formulated by Erwin Schrödinger in 1926, provides a powerful differential equation that governs the time evolution of this wave function. There are two fundamental forms of the Schrödinger equation: the time-dependent Schrödinger equation (TDSE) and the time-independent Schrödinger equation (TISE), each serving distinct purposes in quantum mechanics.

The time-dependent Schrödinger equation is expressed as:

$$i\hbar \frac{\partial}{\partial t} \psi(\mathbf{r}, t) = \hat{H} \psi(\mathbf{r}, t)$$

where i is the imaginary unit, $\hbar$ is the reduced Planck's constant, $\psi(\mathbf{r}, t)$ is the wave function depending on position $\mathbf{r}$ and time t, and $\hat{H}$ is the Hamiltonian operator comprising the total energy of the system. The TDSE describes how the state of a quantum system changes over time, essential for understanding dynamic processes in molecular and atomic systems.

For many practical applications, particularly in quantum chemistry, we focus on systems in a stationary state where the potential does not change with time. Here, the TISE suffices, represented as:

$$\hat{H} \psi(\mathbf{r}) = E \psi(\mathbf{r})$$

In this setting, $\hat{H}$ represents the energy operator, and E denotes the energy eigenvalues associated with each stationary state. Solving the TISE yields the quantized energy levels, leading to insightful revelations about a system's properties and behavior.

The Hamiltonian operator in the Schrödinger equation often partitions into kinetic and potential energy terms. For a single non-relativistic particle in a potential V, the Hamiltonian is expressed as:

$$\hat{H} = -\frac{\hbar^2}{2m} \nabla^2 + V(\mathbf{r})$$

where ∇^2 is the Laplacian operator indicating the sum of second spatial derivatives, and m is the particle's mass. The kinetic energy term captivates the wave-like properties of quantum particles, while the potential energy term $V(\mathbf{r})$ sets constraints reflecting interactions with external fields or forces.

A canonical example illustrating these principles is the quantum harmonic oscillator, a system where a particle experiences a restoring force proportional to its displacement x, represented as:

$$V(x) = \frac{1}{2}m\omega^2 x^2$$

where ω is the angular frequency. The solution of the Schrödinger equation in this context produces discrete energy levels:

$$E_n = \hbar\omega\left(n + \frac{1}{2}\right), \quad n = 0, 1, 2, \ldots$$

reflecting the remarkable nature of quantization, further stressing that even in its lowest energy state ($n = 0$), the system retains a non-zero energy ($\frac{1}{2}\hbar\omega$), known as zero-point energy.

Further insight into the Schrödinger equation's implications is observed in the particle-in-a-box model, a paradigm illustrating the quantization of a particle confined to an infinitely deep potential well. Considering a one-dimensional box of length L, the potential $V(x)$ is zero within the box and infinite outside. The solutions to the Schrödinger equation in this model produce wave functions:

$$\psi_n(x) = \sqrt{\frac{2}{L}} \sin\left(\frac{n\pi x}{L}\right)$$

and corresponding energy levels:

$$E_n = \frac{n^2 \pi^2 \hbar^2}{2mL^2}, \quad n = 1, 2, 3, \ldots$$

These outcomes capture the essence of quantization—energy levels take on discrete values dictated by quantum numbers, exhibiting close adherence to particle-wave duality in a confined system.

Molecule-specific insights arise when approaching multi-particle systems, where more complex potentials and interactions necessitate numerical solutions or approximation methods due to the mathematical complexity. Here, the electronic structure of atoms and molecules is the primary focus, obtained

2.2. THE SCHRÖDINGER EQUATION

by solving the Schrödinger equation for multiple electrons interacting in an electrostatic field produced by nuclei.

This leads to notable applications, including the hydrogen atom problem, where the Schrödinger equation admits exact solutions. For the hydrogen atom, where a single electron orbits a stationary proton, the time-independent Schrödinger equation in spherical coordinates manifests as a product of radial and angular components. The solutions to the radial equation reveal the discrete energy levels:

$$E_n = -\frac{13.6\,\text{eV}}{n^2}, \quad n = 1, 2, 3, \ldots$$

mirroring observed spectral lines in the hydrogen emission spectrum, substantiating the validity of quantum mechanics.

Beyond hydrogen, solving the Schrödinger equation for atoms with more electrons demands computational techniques like Hartree-Fock and Density Functional Theory. These methods simplify the many-body problem by considering electron interactions approximatively or through electron density functionals, providing a practical means for calculating atomistic and molecular properties across diverse systems.

Utilizing these quantum mechanical frameworks extends further into chemistry, enabling the exploration of chemical reaction dynamics, molecular interactions, and the prediction of properties like stability, reactivity, and spectroscopic signatures. Each outcome derives from solving or approximating solutions to the Schrödinger equation, bridging the microscopic quantum world with observable macroscopic phenomena.

In essence, the Schrödinger equation enriches our understanding by offering a unified description of a wide range of physical scenarios, underpinning the diversity and depth of phenomena we observe in chemistry, physics, and beyond. Its impact continues to resonate, as it not only elucidates the fundamental nature of matter but also drives technological advancements and philosophical discourse, reinforcing its significance within and outside the scientific community.

2.3 Quantum Mechanical Models of the Atom

The evolution of atomic models showcases the convergence of empirical observations with theoretical advancements, culminating in the quantum mechanical model that offers an intricate depiction of atomic structure. Early understandings of atomic models were largely descriptive, gradually yielding to more sophisticated representations founded upon quantum mechanics. This section delves into these historical underpinnings and the modern quantum mechanical model, emphasizing the conceptual and mathematical innovations that articulate the atom's intricate architecture.

The conceptual evolution begins with the Bohr model, an advancement over the classical Rutherford model that proposed electrons in fixed orbits around the atomic nucleus. Introduced by Niels Bohr in 1913, this model integrated early quantum ideas to address inadequacies in classical physics, particularly the problem of atomic stability and spectral emissions. In the Bohr model, electrons occupy quantized orbit levels around the nucleus, with energy levels calculated as:

$$E_n = -\frac{Z^2 R_H}{n^2}$$

where Z is the atomic number, R_H is the Rydberg constant, and n is the principal quantum number. These values account for spectral lines observed in hydrogen, specifically the Balmer series, by positing electron transitions between fixed energy levels.

Despite its success in explaining hydrogen spectra, the Bohr model faced limitations when extended to multi-electron systems. To address this, advancements in quantum theory guided the development of the quantum mechanical model of the atom, predicated upon solving the Schrödinger equation. This shift marked a transition from orbits to orbitals—mathematical functions describing regions of probable electron presence.

Orbitals represent solutions to the Schrödinger equation characterized by sets of quantum numbers: principal (n), azimuthal (l), magnetic (m_l), and spin (m_s). The principal quantum number, n, determines the size and energy, whereas the azimuthal quantum number, l, indicates orbital shape. The magnetic quantum number, m_l, influences orbital orientation, and the spin quantum number, m_s, accounts for intrinsic electron spin.

2.3. QUANTUM MECHANICAL MODELS OF THE ATOM

The quantum mechanical model's nuances manifest vividly in the hydrogen atom, where wave functions are expressed in spherical coordinates as $\psi(r, \theta, \phi)$, comprised of radial and angular components. Solutions yield distinct wave functions or orbitals, represented as s, p, d, and f, each with characteristic shapes and nodal structures. The electron probability distribution for these orbitals defines the atom's electron cloud, demystifying electron distributions and interactions in space.

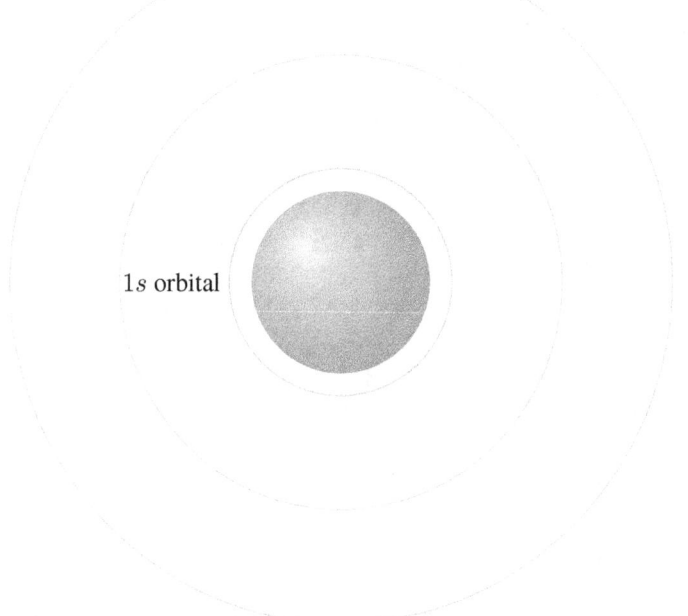

1s orbital

The quantum mechanical model proposes electron configurations that align with elemental periodicity. Electrons fill orbitals according to the Aufbau principle, Pauli exclusion principle, and Hund's rule. The Aufbau principle suggests electron filling starts from lower to higher energy orbitals, constrained by the Pauli exclusion principle, which prohibits identical sets of quantum numbers among electrons, ensuring distinct electron states. Hund's rule optimizes electron arrangement by maximizing unpaired spins in degenerate orbitals, minimizing repulsive interactions.

Consider the helium atom, where both electrons inhabit the 1s orbital—the simplest multi-electron configuration—rooted in minimizing system energy. Progressing through the periodic table ratifies the orbital filling order, substantiating predicted valencies, chemical reactivity, and atomic spectra.

When addressing transition elements, d-orbital involvement and electron correlations render the picture intricate, explaining phenomena like variable oxidation states and magnetic behaviors. In f-block elements, the f orbitals add further complexity, influencing lanthanide contraction and related properties.

While this model applies robustly to most elements, the challenges posed by many-electron systems necessitate numerical methods and approximations. Approaches like Hartree-Fock and Density Functional Theory deconstruct complex electron-electron interactions and calculate properties with high precision, generating valuable insights into chemical bonding, electronic transitions, and molecular geometries.

Modern computational techniques grant higher accuracy by employing basis sets and electron correlation methods, advancing predictions pertinent to materials design, catalytic processes, and drug development. As such, the quantum mechanical model's role in advancing chemistry and materials science remains essential and continuously evolving with technological maturation.

The quantum mechanical model transcends quantitative analysis, embodying a conceptual paradigm shift that reshaped scientific inquiry into atomic and subatomic phenomena. The interplay of wave-particle duality, uncertainty, and quantization characterizes this shift, influencing adjacent disciplines and fostering explorations into quantum computing, nuclear physics, and cosmology.

The envisaging of atomic structure as probabilistic rather than deterministic heralds the era of quantum mechanics, where logical determinism yields to stochasticity, and the fabric of matter is elucidated through intricate possibilities rather than certainties. This profound framework perpetuates exploration and innovation, sustaining an inquisitive odyssey through the depths of atomic complexity.

2.4 Atomic Orbitals and Electron Configurations

Atomic orbitals represent fundamental concepts in the quantum mechanical model of the atom, providing a detailed depiction of the spatial distribution and energy of electrons in an atom. An understanding of atomic orbitals forms

2.4. ATOMIC ORBITALS AND ELECTRON CONFIGURATIONS

the basis for predicting the chemical behavior and properties of elements, as the distribution of electrons within these orbitals determines interaction potential and chemical bonding. In this section, we explore the shapes, orientations, and energies of atomic orbitals, and how electrons configure themselves within these orbitals according to well-established quantum principles.

The Schrödinger equation is solved to yield wave functions, ψ, which describe the state of electrons in atoms. These wave functions, when squared, provide a probability density function for the location of an electron within an atom. The solutions are characterized by distinct quantum numbers that form the quantum framework essential to understanding atomic structure. The principal quantum number, n, indicates the energy level and size of the orbital, while the azimuthal quantum number, l, determines the shape. The magnetic quantum number, m_l, describes the orientation of the orbital in space, and the spin quantum number, m_s, differentiates the intrinsic angular momentum of electrons.

The principal quantum number, n, is a positive integer starting from 1. It predominantly governs the size and energy of the orbital—the greater the value of n, the larger and more energetic the orbital. Thus, orbitals with higher n values are physically bigger and further from the nucleus. The azimuthal quantum number, l, takes on integer values ranging from 0 to $n - 1$. These values correspond to different orbital shapes commonly labeled as s, p, d, and f. For example, $l = 0$ characterizes s orbitals, $l = 1$ characterizes p orbitals, $l = 2$ corresponds to d orbitals, and so on.

The spherical shape of s orbitals allows for equal probability distributions around the nucleus whereas p orbitals, bearing dumbbell shapes, display lobes directed along specific axes in three-dimensional space. Their orientation depends on the magnetic quantum number (m_l), which ranges from $-l$ to $+l$, affecting orbital orientation thus: p orbitals ($l = 1$) can have orientations corresponding to $m_l = -1, 0, 1$, labeled as p_x, p_y, and p_z. Similarly, d orbitals ($l = 2$) present more complex shapes and orientations, involving multiple lobes and nodal planes, critical for transition metals where d-orbital occupancy influences chemical interaction significantly.

The spin quantum number, m_s, introduces a critical dimension by explaining electron spin, an intrinsic form of angular momentum. It takes values of $+\frac{1}{2}$ or $-\frac{1}{2}$ corresponding to two possible spin states, commonly termed "spin-up" and "spin-down." The spin quantum number is paramount because of the Pauli exclusion principle, which asserts that no two electrons in an atom can have identical sets of all four quantum numbers, leading to the necessity for electron spin differentiation within orbitals.

Configurations of electrons around an atom's nucleus are neatly expressed through electron configurations, which reflect the increasing energy and occupancy of orbitals following the Aufbau principle. This principle guides the process of filling atomic orbitals starting from lower to higher energy states. The typical sequence showcases energy sublevels filling in the following order: 1s, 2s, 2p, 3s, 3p, 4s, 3d, 4p, 5s, with d orbitals filling subsequent to the intervening s sublevels.

Principal Level	Subshells
$n = 1$	s
$n = 2$	s, p
$n = 3$	s, p, d
$n = 4$	s, p, d, f

For instance, the electron configuration for hydrogen, encompassing a single electron, is denoted as $1s^1$, while helium, with two electrons in the $1s$ orbital, is denoted by $1s^2$. Progressing to neon, the configuration completes its second shell: $1s^2\ 2s^2\ 2p^6$.

A detailed understanding of electron configurations goes beyond mere occupational patterns. It extends into chemical periodicity where the configuration underlies the recurring characteristics and behavior observed in elements of the periodic table. Electron configurations elucidate the filling of s, p, d, and f blocks, offering profound insight into the periodic table's arrangement and elemental reactivity, among other properties, aligning predictive capacities in bonding tendencies and chemical reactions.

Considerations such as Hund's rule, where electrons occupy degenerate orbitals singly before pairing, endorse electron distribution that minimizes electron-electron repulsions and stabilizes atom configuration. This rule explains cases where energy considerations lead to configurations like the nitrogen atom, represented as $1s^2\ 2s^2\ 2p^3$, with each of the three p orbitals singly occupied.

Exploring transition metals requires appreciating nuances, as their d-orbital occupancy permits shared characteristics between different elements, promoting phenomena such as variable oxidation states. These properties, owing much to the d-block's positional versatility and energetic proximity to s-orbitals, fashion unique chemical behaviors leading to complex coordination chemistry exemplified through known transition ions.

The f-block elements, lanthanides, and actinides reveal further complexities, given their similar electron filling sequences—their obscure role in periodic structure emphasizes thematic periodicity and nuanced understanding of electron-electron interactions in determining characteristics.

Beyond theoretical predictions, this orbital-based understanding of electron configurations significantly impacts experimental chemistry, manifesting in empirical observations. For instance, atomic and electronic spectra analyze light absorption and emission patterns to infer electron transitions among orbital energy levels, often depicted as line spectra specific to elements.

Atomic orbitals and electron configurations are cornerstones of modern atomic theory. From simple hydrogen to complex heavy elements, understanding how electrons populate atomic orbitals elucidates the intricate and beautiful complexity of the periodic table, forming a bridge between theoretical principles and observable chemical properties that make up the fundamental aspects of chemistry and material sciences. These concepts provide the scaffolding necessary for theorizing about matter, crafting new materials, and anticipating chemical reactions crucial for scientific advancement and innovation.

2.5 Quantum Numbers and Atomic Structure

Quantum numbers are integral to the quantum mechanical model, facilitating a systematic understanding of atomic structure. They arise from solutions to the Schrödinger equation for electrons in an atom, embodying essential degrees of freedom to describe fully an electron's spatial configuration and energetic state. These numbers—principal, azimuthal, magnetic, and spin—enrich our comprehension of atomic characteristics, dictating electron configurations and elemental chemical behaviors.

The principal quantum number, n, is the foundation of the hierarchy of quantum numbers. It is a positive integer value ($n = 1, 2, 3, \ldots$) and heralds the electron's energy level or shell, a notion carrying both size and energy implications. Electrons in orbitals of higher n levels reside further from the nucleus and are associated with higher energy states. Therefore, it conceptually aligns with the Bohr model of quantized electron orbits, reflecting the energy needed to remove an electron from an atom—a larger n denotes a lesser electrostatic attraction to the nucleus.

Upon delving deeper into atomic structure, we encounter the azimuthal quantum number, l, also known as the angular momentum quantum number. The values of l range from 0 to $n-1$, and this quantum number intrinsically determines the orbital's shape, influencing nodal patterns and the angular distribution of electrons. Corresponding to specific orbitals, $l = 0$ defines s orbitals, $l = 1$ defines p orbitals, $l = 2$ equates to d orbitals, and $l = 3$ corresponds to f orbitals. These diverse orbital shapes underscore electron density contours shaping chemical interactions and bonding geometries.

Subshell	s	p	d	f
l	0	1	2	3

The magnetic quantum number, m_l, allows for greater discernment by describing an orbital's orientation in three-dimensional space. The range of m_l encapsulates integers from $-l$ to $+l$, accommodating multiple spatial orientations of a given shape, corresponding to orbital degeneracy under the influence of external fields. Within a p orbital ($l = 1$), for instance, m_l takes values of $-1, 0, 1$, designating orientations corresponding to p_x, p_y, and p_z. Introducing external fields can resolve these degeneracies, splitting energy levels through phenomena like the Zeeman effect where applied magnetic fields modulate atomic spectra, offering insights into electronic interactions.

Completing our quantum number set is the spin quantum number, m_s, which posits two possible values $(+\frac{1}{2}, -\frac{1}{2})$, explaining the binary nature of electron spin. Electron spin mimics intrinsic angular momentum, playing a crucial role in addressing magnetic properties at the atomic and molecular levels. The pivotal Pauli exclusion principle emerges to stipulate that no two electrons in a given atom can possess an identical set of quantum numbers, enforcing unique states and ensuring atomic integrity. This principle is instrumental in understanding electron configuration and stability, mandating paired spins to enforce fulfillment of occupied orbitals.

Quantum numbers, indeed, shape the atomic landscape, governing the electron configuration and chemical reactivity. The presence of these quantum indices dictates permissible configurations, accommodating electron filling in a systematic sequence as dictated by the Aufbau principle. Here, electron configurations unfold, sequentially filling lower-energy orbitals preceding higher-energy alternatives. This principle reinforces periodic table architecture, underpinning elemental periodicity witnessed across properties and reactivity trends throughout the periodic table.

2.5. QUANTUM NUMBERS AND ATOMIC STRUCTURE

Examining a practical example, consider carbon, whose electron configuration is $1s^2 2s^2 2p^2$. In carbon's ground state, governed by Hund's rule—stating that electrons will singly occupy degenerate orbitals before pairing for minimal electron-electron repulsion—we find paired electrons in the 1s and 2s orbitals while occupying two separate 2p orbitals. This distribution facilitates chemical flexibility, protecting carbon's versatility and enabling hybridization, crucial for accommodating tetrahedral geometry and sp^3 hybrid bonds seen in countless organic compounds.

An impactful illustration lies within transition metals, where d orbitals take center stage. For instance, iron with an electron configuration of $[Ar] 3d^6 4s^2$, presents unpaired d electrons—an explanation for its ferromagnetic properties. The electron interchange among d orbitals and their proximity to s-orbitals allow bridging between discrete energetics, facilitating variable oxidation states that exemplify ionic versatility in catalytic processes and complex formations.

In the lanthanide and actinide series, quantum mechanics explicates properties driving these f-block elements, where the role of f orbitals becomes significant. The incorporation of these orbitals connects directly to lanthanide contraction and chemical uniqueness, impacting atomic radii and reactivity. With electron configurations challenging simplistic periodic relationships, intensive study resolves such intriguing elements into coherent, predictable chemical characterizations.

The integration of quantum numbers with atomic structure finds extension in spectroscopy, providing indirect evidence about energy level separations and electron transitions. Atomic absorption and emission spectra dissect light interaction with atoms through electronic transitions among energy levels, describing observable line spectra or spectral series unique to elemental species. Here, the Balmer and Lyman series conceptually illustrate hydrogen's excited electron returning to lower states, revealing important quantum leaps that reflect hydrogen atom structure.

The resonance of quantum numbers with atomic structure gestures towards profound consequences straddling chemistry and physics. Impacting beyond theoretical bounds, these concepts have engineered tremendous advancements begetting modern atomic clocks, facets of quantum computing, and quantum cryptography—pushing theoretical constructs into transformative real-world applications.

Quantum numbers enfold the complexities of atomic extent and subtend a coherent architectural marvel inside the atom. They depict not mere numerological representations but formulate a nuanced, deeply-intertwined quantum vista, furthering atomic theory's advance and bestowing a refined harmonization of theory with empirical vistas. These numbers offer a map—carefully charting the quantum and discreetly guiding physicists and chemists alike through the maze of atomic behavior intricately woven into the universe's tapestry.

2.6 Approximation Methods in Quantum Chemistry

Quantum chemistry chiefly relies upon the principles of quantum mechanics to elucidate the structure and behavior of molecules. However, the inherent complexities associated with many-electron systems make exact solutions impractical for all but the simplest atoms and ions. Consequently, approximation methods are employed extensively to simplify and solve the Schrödinger equation for complex molecular systems. These techniques pivotal in quantum chemistry include the Hartree-Fock method, post-Hartree-Fock techniques, and Density Functional Theory (DFT).

The Hartree-Fock method represents one of the primary approximation approaches within quantum chemistry. It is an iterative procedure designed to approximate the wave function and energy of a quantum many-body system by considering the average effect of electron-electron interactions. Fundamentally, it assumes that the exact, correlated many-electron wave function can be approximated by a single Slater determinant of spin orbitals, ensuring compliance with the Pauli exclusion principle and electron indistinguishability.

The Hartree-Fock approximation employs a mean-field approach where each electron moves in an average potential field created by the nuclei and the other electrons. This decouples the multi-electron problem into a set of single-electron equations, known as the Hartree-Fock equations. The resultant self-consistent field (SCF) method iteratively refines the potential until convergence is achieved:

$$\hat{F}\psi_i = \varepsilon_i \psi_i$$

where $\hat{F}$ is the Fock operator, ψ_i are the molecular orbitals, and ε_i are the orbital energies. Despite its efficacy in generating qualitative insights and the

2.6. APPROXIMATION METHODS IN QUANTUM CHEMISTRY

foundational basis it provides, Hartree-Fock neglects electron correlation—a limitation particularly for systems where electron correlation significantly influences chemical behavior, such as in transition metal complexes or bond dissociation processes.

To address these shortcomings, post-Hartree-Fock methods have been developed, which incorporate electron correlation effects systematically. Among these are Configuration Interaction (CI), Coupled Cluster (CC) methods, and Møller-Plesset perturbation theory.

- **Configuration Interaction (CI)** aims to improve upon Hartree-Fock by including a linear combination of excited Slater determinants in the wave function expansion:

$$\Psi = c_0 \Phi_0 + \sum_i c_i \Phi_i$$

- where Φ_0 is the Hartree-Fock reference determinant, and Φ_i represents excited configurations with c_i as coefficients determined by diagonalizing the Hamiltonian within the chosen determinant space. Although CI potentially reaches exact solutions within the given basis, computational costs rise steeply with system size and electron number, rendering it infeasible for very large systems.

- **Coupled Cluster (CC) methods** provide another robust framework by employing an exponential ansatz to account for electron correlations—a more compact and systematically improvable representation than CI. The coupled cluster wave function assumes the form:

$$\Psi = e^{\hat{T}} \Phi_0$$

- where $\hat{T}$ is the cluster operator comprising excitation terms ($T_1, T_2, \ldots$). Notably, the CC method, with single and double excitations (CCSD), or with triples (CCSDT), balances accuracy with computational practicability, offering highly accurate solutions for electronic correlation without exhaustive computational demands, albeit with increased sophistication required for implementation.

- **Møller-Plesset Perturbation Theory (MP2)** is a popular method when seeking computational efficiency, leveraging Rayleigh-Schrödinger perturbation theory in reforming electron correlation calculations beyond Hartree-Fock. Specifically, MP2 examines the energy correction from the second-order perturbation, contouring a balance between computational expenditure and accuracy. It serves as a valuable component in quantum chemical calculations, enabling meaningful approximations while preserving manageable computational requirements.

The **Density Functional Theory (DFT)** represents an entirely different approach by substituting wave function complexities with which electron probability densities are central. DFT reformulates quantum mechanical problems in terms of electron density instead of many-body wave functions, with the Hohenberg-Kohn theorems providing a rigorous foundation, stating that the ground-state properties of a many-electron system are uniquely determined by its electron density $\rho(\mathbf{r})$. The celebrated Kohn-Sham formulation recasts interacting electron problems into an equivalent non-interacting system:

$$\left(-\frac{\hbar^2}{2m}\nabla^2 + V_{\text{eff}}(\mathbf{r})\right)\psi_i(\mathbf{r}) = \varepsilon_i\psi_i(\mathbf{r})$$

where $V_{\text{eff}}(\mathbf{r})$ is the effective potential, encompassing external potential, Hartree potential, and exchange-correlation potential. The latter remains central to DFT's success, with approximations like Local Density Approximation (LDA), Generalized Gradient Approximation (GGA), and hybrid functionals forming its lifeblood, offering pragmatic and efficient electron correlation descriptions across many chemical systems.

Within the remit of practical computational chemistry, approximation methods empower chemists to explore a plethora of systems, from small biomolecules to extensive condensed matter applications, including catalysis, materials design, and nanoscale configurations. They facilitate the description of bonding nuances, reaction pathways, spectral properties, and dynamic behaviors at the quantum level, significantly raising the analytical and predictive capabilities of theoretical chemical research.

The chosen method invariably depends on the complexity of the system studied, the accuracy needed, and computational resources available. Hartree-Fock and DFT dominate larger systems where electron correlation plays less critical roles or where hybrid-functional accuracy suffices. Conversely, post-

Hartree-Fock methods like CI and CC are preferred where precision warrants addressing non-covalent interactions, bond breaking, or providing benchmarking studies to validate lesser computational methods.

Approximation methods constitute the backbone of quantum chemistry, reconceptualizing the treatment of electron-electron interactions to demystify intricate molecular systems. As methodological advancements and computational capabilities grow, these approximations continue to advance, fueling deeper insights and innovative exploration across theoretical and applied domains in chemistry and materials sciences. Ultimately, these methods bridge theoretical aspirations with tangible experimental inquiries, unraveling the quantum mechanical pathways within chemical phenomena and beyond.

2.7 Applications of Quantum Chemistry in Spectroscopy

Quantum chemistry provides indispensable insights into the electronic structure and behavior of atoms and molecules, which are pivotal for interpreting spectroscopic data. Spectroscopy, encompassing a broad array of techniques, relies on the interaction of electromagnetic radiation with matter, providing a window into the quantum world and the transitions between energy levels within atoms and molecules. Through quantum chemistry, the connection between spectral lines and molecular characteristics is demystified, allowing for profound interpretation and application across various scientific domains.

The quantum mechanical model of atoms and molecules establishes a firm foundation for understanding spectroscopy. At its core, spectroscopy examines transitions from one quantum state to another, often involving changes in the electronic, vibrational, or rotational energy states of a molecule. Each transition corresponds to the absorption or emission of radiation at specific frequencies, forming the basis for the identification and analysis of substances.

One of the simplest and most illustrative applications of quantum chemistry in spectroscopy is seen in atomic spectroscopy, which investigates the electromagnetic radiation absorbed and emitted by atoms. It delivers insights through line spectra, which directly result from electronic transitions between quantized energy levels within atoms. The hydrogen atom offers a classic example, where quantum mechanical solutions to the Schrödinger equation elucidate the Balmer series in its emission spectrum, highlighting transitions as electrons move between higher energy levels and the second energy level:

$$E_n = -\frac{13.6\,\text{eV}}{n^2}$$

where the observed spectral lines equate to specific electronic transitions visible as distinct colors in the hydrogen spectrum under excitation. Through quantum chemistry, the precise nature of these transitions, their selection rules, and their spectral properties—such as line widths and intensities—are interpreted effectively.

Transitioning from atoms to molecules, molecular spectroscopy delves into the electronic, vibrational, and rotational states of molecules. Each molecular motion type corresponds to distinctive regions in the electromagnetic spectrum. Electronic transitions, typically in the ultraviolet and visible regions, necessitate an understanding of molecular orbitals and electron configuration, where valence electrons transition between bound molecular orbitals, characterized by differences in molecular energy:

$$\Delta E = E_{\text{higher}} - E_{\text{lower}}$$

This foundational level understanding paves the way for techniques like UV-Vis spectroscopy, pivotal in determining conjugation in organic compounds. Quantum chemical methods, particularly time-dependent DFT (TD-DFT), enable precise computation of excitation energies, simulating absorption spectra, and thereby elucidating the structural and electronic attributes of molecules.

The interplay of vibrational and rotational transitions finds its sphere within infrared (IR) and Raman spectroscopy. IR spectroscopy is concerned with vibrational transitions corresponding to a molecule's internuclear bond lengths and angles, reported as specific frequencies where molecules absorb radiative energy. The quantum mechanical model describes this harmonic oscillator-like behavior mathematically, where vibrational frequencies are determined using relations involving bond force constants and atomic masses.

$$\nu = \frac{1}{2\pi}\sqrt{\frac{k}{\mu}}$$

where k is the force constant, and μ is the reduced mass of interacting atoms. Quantum chemistry supports the interpretation of IR spectra by detailing nor-

2.7. APPLICATIONS OF QUANTUM CHEMISTRY IN SPECTROSCOPY

mal modes of vibration and correlating these observables with molecular structure and environment, enhancing understanding of functional groups, bonding patterns, and structural dynamics.

Raman spectroscopy, another embodiment of vibrational analysis, exploits inelastic scattering of light, providing complementary information to IR spectroscopy. Quantum mechanical treatment of the Raman effect involves induced polarization of molecules by incident photons, transitioning a molecule from an initial to a virtual energy state and scattering photons at shifted frequencies indicative of vibrational modes.

In nuclear magnetic resonance (NMR) spectroscopy, quantum chemistry provides profound insights into nuclear environments through magnetic interactions at the atomic nucleus level. NMR exploits quantum mechanics by manipulating nuclear spin states in an external magnetic field, inducing transitions whose frequencies relate directly to chemical shifts, coupling constants, and relaxation times. Leveraging quantum chemical calculations, one can simulate and predict NMR spectra, interpreting electronic environments, molecular conformation, and dynamic processes within complex molecular systems, pivotal in elucidating large biomolecules like proteins or nucleic acids.

Advanced quantum chemical techniques also facilitate electron spin resonance (ESR) and Mössbauer spectroscopy, offering precise understanding of unpaired electron systems and nuclear interactions in solids. Quantum mechanics nurtures these spectroscopies, providing frameworks to simulate and interpret spectra, deducing structural and electronic information critical in fields like materials science, catalysis, and forensic analysis.

Mass spectrometry, though not a traditional spectroscopic technique, complements quantum chemical approaches by analyzing ionized fragments based on mass-to-charge ratios. Here, insights from quantum chemistry assist in recognizing fragmentation patterns, elucidating structural motifs, and enhancing molecular identification efficiency.

Quantum chemistry and spectroscopy jointly offer tools essential for deducing fundamental molecular properties, including bond lengths, angles, electronic distributions, and energy landscapes. Through computational methods like ab initio approaches and DFT, researchers obtain theoretical spectra which when compared with experimental spectra, dramatically enhance interpretive accuracy and chemical understanding.

The applications extend into examining reaction dynamics, where time-resolved spectroscopic techniques disclose temporal evolution of intermediates and transition states, furthering the grasp on mechanistic and kinetic aspects of chemical processes.

Quantum chemistry underpins the profound analysis and interpretation of spectroscopic data, revealing through the lens of quantum mechanics an intricate portrait of molecular systems. Spectroscopy, complemented by quantum insights, transcends mere observation, enabling precise characterization, interaction elucidation, and the prediction of chemical behavior, crucial for scientific advancement across disciplines embracing chemistry, biology, physics, and material sciences. This integration illuminates the path to demystifying the complexities of the molecular universe, guiding future exploration and technological innovation in the ever-evolving realm of science.

Chapter 3

Molecular Bonding and Structure

The intricate nature of molecular bonding and structure underpins the diversity and functionality of chemical substances, where distinct types of bonds manifest the dynamic interactions between atoms. This chapter explores the theoretical foundations governing bond formation, emphasizing the concepts of valence bond and molecular orbital theories. By analyzing hybridization and molecular geometry, the spatial arrangements of atoms within molecules are revealed, influencing polarity and intermolecular forces. With a focus on resonance and electron delocalization, the effect of these factors on molecular stability and reactivity is articulated, providing a comprehensive perspective on how atomic connectivity dictates the chemical and physical properties of matter.

3.1 Types of Chemical Bonds

Chemical bonds are the connections that hold atoms together, forming molecules and compounds. Understanding the nature of these bonds is essential for comprehending the structure and properties of matter. The main types of chemical bonds are ionic, covalent, and metallic bonds, each characterized by the specific interaction mechanisms between atoms.

Ionic Bonds

Ionic bonds are formed between atoms that have significantly different electronegativities, usually between metals and non-metals. In this case, one atom donates one or more electrons to another, resulting in the formation of ions. The atom that loses electrons becomes a positively charged ion, or cation, while the atom that gains electrons becomes a negatively charged ion, or anion. The electrostatic attraction between these oppositely charged ions constitutes the ionic bond.

For example, in sodium chloride, NaCl, sodium (Na) donates an electron to chlorine (Cl). Sodium becomes Na^+ and chlorine becomes Cl^-, forming a stable ionic compound. The strength of ionic bonds is influenced by the charges on the ions and the size of the ions. Greater charges result in stronger electrostatic attraction, while smaller ions allow for closer proximity, also increasing the bond strength.

Lattice energy is a key concept in understanding ionic bonds. It is the energy released when ions come together to form a crystalline lattice. Lattice energy can be estimated using the Born-Haber cycle, a thermodynamic cycle that applies Hess's law to ionic compounds. This cycle considers the energy changes associated with the formation of an ionic solid from its constituent elements.

Covalent Bonds

Covalent bonds arise when atoms share pairs of electrons, typically between non-metal atoms with similar electronegativities. The shared electrons allow each atom to attain a stable electronic configuration, resembling that of a noble gas. Covalent bonding results in the formation of molecules, ranging from simple diatomic molecules like hydrogen, H_2, to complex organic compounds.

The bond strength in covalent bonds is dependent on the number of shared electron pairs. Single bonds involve one shared pair, double bonds involve two shared pairs, and triple bonds involve three shared pairs of electrons. Consequently, triple bonds are generally stronger and shorter than double and single bonds. For example, in nitrogen, N_2, the two nitrogen atoms are connected by a triple bond, which accounts for the molecule's exceptional stability.

Polarity is another significant feature of covalent bonds. When atoms with different electronegativities form a covalent bond, the electrons are not shared equally, resulting in a polar covalent bond. The more electronegative atom attracts the shared electrons more strongly, acquiring a partial negative charge, while the other atom acquires a partial positive charge. The measure of bond polarity is quantified by dipole moment, which reflects the degree of charge separation within the molecule.

Metallic Bonds

Metallic bonds occur between metal atoms, where the valence electrons are not associated with any specific atom but are instead delocalized across the entire structure. This electron 'sea' allows for the electrons to move freely, facilitating the unique properties of metals such as electrical conductivity, malleability, and ductility.

In metallic bonding, positively charged metal ions are held together by their attraction to the delocalized electrons. This model explains why metals can conduct electricity; electrons can move through the metal to conduct an electric current. Furthermore, the presence of delocalized electrons enables metals to be hammered into sheets or drawn into wires without breaking the metallic bonds, due to the non-directional nature of the bond.

The strength of metallic bonds depends on the charge of the metal ions and the sea of electrons. Transition metals, with their ability to have variable oxidation states and partially filled d orbitals, often exhibit stronger metallic bonding, resulting in higher melting and boiling points compared to alkali metals.

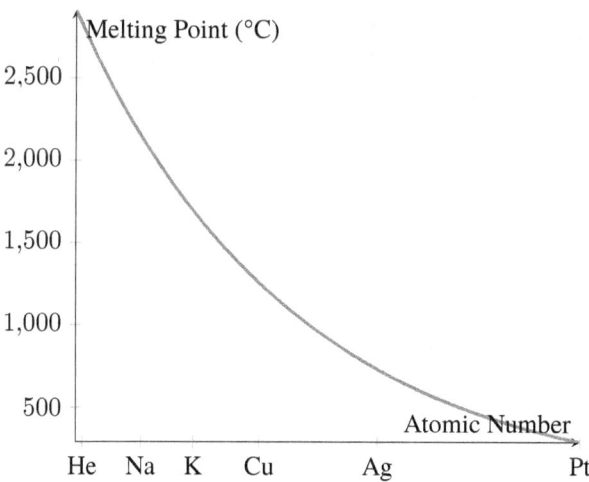

Overall, the three primary types of chemical bonds - ionic, covalent, and metallic - each contribute unique characteristics to the compounds and elements they form. Ionic bonds produce high melting and boiling points and electrical conductivity in the molten state due to the robust electrostatic forces between ions. Covalent bonds provide structural stability and chemical specificity through electron sharing, while metallic bonds impart the distinct and utilitarian properties of metals.

Analyzing the interaction at the subatomic level in each bond type enables chemists to predict the behavior and properties of diverse chemical systems. This understanding forms the basis for advanced exploration in chemical synthesis, materials science, biochemistry, and other related fields, where manipulation of atomic bonds leads to novel molecular architectures and functional materials.

3.2 Valence Bond Theory

Valence bond (VB) theory is an essential concept for understanding the nature of chemical bonds within molecules. This theoretical framework describes how atomic orbitals of individual atoms overlap to form chemical bonds, providing insight into the directionality and strength of bonds within molecules.

Valence bond theory is grounded in the quantum mechanical model, which describes electrons in terms of probabilities and spatial distributions, rather than fixed orbits. Within this paradigm, valence electrons are those involved in bond formation, and they reside in atomic orbitals—regions around a nucleus where there is a high probability of finding electrons.

One of the conceptual cornerstones of valence bond theory is the idea of orbital hybridization. In its simplest form, hybridization involves the mixing of atomic orbitals to form new hybrid orbitals. These hybrid orbitals have shapes and energies that are favorable for maximizing the overlap with orbitals from other atoms, enhancing bonding interactions.

Orbital Overlap and Bonding

In valence bond theory, a covalent bond is formed when the atomic orbitals of two atoms overlap, allowing their electrons to be shared. The strength of the bond is proportional to the extent of orbital overlap. Greater overlap leads to stronger bonds. For example, the formation of a hydrogen molecule (H_2) occurs when the 1s orbitals of two hydrogen atoms overlap, sharing a pair of electrons and forming a stable bond.

3.2. VALENCE BOND THEORY

Valence bond theory attributes directional properties to certain bonds due to the specific geometries of overlapping orbitals. For instance, in the case of CH_4, the tetrahedral geometry observed can be explained through the hybridization of one 2s and three 2p orbitals from the carbon atom, resulting in four identical sp^3 hybrid orbitals. Each of these hybrid orbitals overlaps with the 1s orbital of a hydrogen atom, creating four equivalent, sigma (σ) bonds, which are characterized by direct overlap of orbitals along the internuclear axis.

Hybridization of Atomic Orbitals

The concept of hybridization is introduced to explain the geometry and bonding of molecules that cannot be described by the simple overlap of atomic orbitals. By considering hybridization, chemists can predict the molecular shapes that correspond to the most energetically favorable configuration of bonding and non-bonding orbitals.

- *sp Hybridization:* This occurs when one s orbital mixes with one p orbital, producing two equivalent sp hybrid orbitals arranged linearly with a bond angle of 180°. This is typically seen in molecules like beryllium chloride ($BeCl_2$).

- *sp^2 Hybridization:* Involves the mixing of one s and two p orbitals, resulting in three sp^2 hybrid orbitals organized in a trigonal planar geometry, with bond angles of 120°. This hybridization is evident in molecules like boron trifluoride (BF_3).

- *sp^3 Hybridization:* Mixing one s and three p orbitals gives four sp^3 hybrid orbitals, arranged in a tetrahedral geometry with bond angles of 109.5°. This configuration is common in methane (CH_4).

- *sp^3d and sp^3d^2 Hybridization:* Seen in molecules with expanded octets, such as phosphorus pentachloride (PCl_5) and sulfur hexafluoride (SF_6), where d orbitals are involved allowing geometries like trigonal bipyramidal and octahedral respectively.

Resonance in Valence Bond Theory

Resonance is a significant phenomenon that extends the valence bond theory to account for certain molecular structures that cannot be described adequately by a single Lewis structure. Resonance structures are a set of two or more valid Lewis structures that define the actual distribution of electrons within a molecule.

A classic example is benzene (C_6H_6), where valence bond theory visualizes the molecule as a combination of resonance structures. In benzene, the carbon atoms are connected by sigma bonds, and the overlapping of adjacent p orbitals results in a delocalized pi electron system, giving rise to the actual structure that is a resonance hybrid of individual structures with alternating single and double bonds.

Limitations of Valence Bond Theory

Though valence bond theory is instrumental in explaining and predicting molecular structure and bonding, it does have limitations. For example, VB theory focuses on localized electron pairs, making it less effective in treating delocalized systems compared to molecular orbital (MO) theory. Moreover, VB theory can be cumbersome in dealing with molecules exhibiting significant resonance or in predicting magnetic properties.

Additionally, while valence bond theory provides insights into the angles and directions of bonds, it does not directly account for bond energies and bond lengths. These properties often require computational techniques or empirical corrections to predict accurately.

Illustrations and Applications

To further elucidate valence bond theory, consider the illustration of molecular geometries through orbital diagrams, which are valuable in visualizing how hybridized orbitals arrange themselves around a central atom. In practical applications, valence bond theory aids in understanding reaction mechanisms, particularly in organic chemistry, where the concept of hybridization facilitates the prediction of molecular reactivity and stereochemistry.

Another practical application of VB theory is in the design and synthesis of coordination compounds, where it explains the bonding in complex ions, accounting for the overlap of metal atomic orbitals with the ligand orbitals. The hybridization concept is also useful in catalysis, where the geometric and electronic properties of catalyst molecules determine their function and efficiency.

3.3. MOLECULAR ORBITAL THEORY

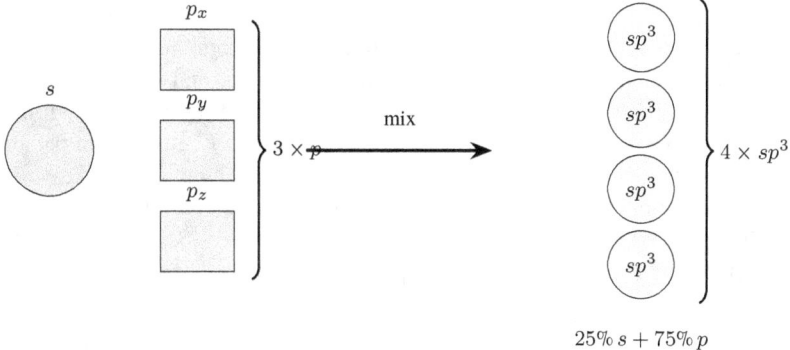

Valence bond theory is a powerful framework for understanding the localized bonding interactions within molecules. While the theory excels in explaining the directional and orbital characteristics of covalent bonds, its limitations point to the need for complementary theories like molecular orbital theory for a complete understanding of molecular bonding. Nevertheless, valence bond theory remains a cornerstone in chemical education and research, providing foundational insights into the nature of chemical bonds.

3.3 Molecular Orbital Theory

Molecular orbital (MO) theory is a more advanced and comprehensive model for understanding the electronic structure of molecules, complementing and extending the valence bond theory. MO theory was developed to address some of the limitations of valence bond theory, particularly in describing the electronic distribution in molecules with delocalized electrons and in explaining magnetic properties.

In molecular orbital theory, atomic orbitals combine to form molecular orbitals that are delocalized over the entire molecule. These molecular orbitals can accommodate the electrons of the molecule, filling according to the principles of quantum mechanics, particularly the Aufbau principle, the Pauli exclusion principle, and Hund's rule. The formation, energy, and occupancy of these molecular orbitals allow for a detailed description of bonding, antibonding, and non-bonding interactions within a molecule.

- **Formation of Molecular Orbitals**

Molecular orbitals are formed by the linear combination of atomic orbitals (LCAO). When atomic orbitals from different atoms combine, they produce molecular orbitals as either bonding or antibonding orbitals. The bonding molecular orbitals result from the constructive interference of atomic orbital wave functions, leading to an increased electron density between the nuclei, which results in lower energy and hence contributes to bond formation. Conversely, antibonding molecular orbitals arise from the destructive interference, characterized by a node between the nuclei, higher energy, and a tendency to weaken or prevent bonding.

For diatomic molecules such as hydrogen (H_2), the 1s atomic orbitals of each hydrogen atom combine to form a sigma (σ) bonding molecular orbital and a sigma star (σ^*) antibonding molecular orbital. The electrons from each hydrogen atom initially fill the lower-energy σ orbital, resulting in a stable covalent bond.

$$H(1s^1) + H(1s^1) \rightarrow H_2 \quad (\sigma, \sigma^*)$$
$$\sigma = \text{bonding molecular orbital}$$
$$\sigma^* = \text{antibonding molecular orbital}$$

- **Classification and Symmetry of Molecular Orbitals**

Molecular orbitals are classified into sigma (σ) and pi (π) molecular orbitals based on their symmetry and orientation relative to the internuclear axis. Sigma orbitals arise when the overlap is symmetrical around the axis connecting two atomic nuclei, allowing for end-to-end overlap of orbitals. Pi molecular orbitals result from the side-to-side overlap of p orbitals above and below the internuclear axis, characteristic of double and triple bonds in molecules like ethylene (C_2H_4) and acetylene (C_2H_2).

The symmetry properties of molecular orbitals are important for understanding chemical reactivity and molecular interactions, as the symmetry dictates how orbitals can combine or interact during chemical reactions. Molecular orbital symmetry also plays a vital role in spectroscopic transitions, as only certain symmetry-allowed transitions can occur.

- **Molecular Orbital Diagrams**

A molecular orbital diagram is a vital tool in MO theory, providing a visual representation of the relative energy levels and electron configurations of the

3.3. MOLECULAR ORBITAL THEORY

molecular orbitals in a molecule. These diagrams illustrate how atomic orbitals from the constituent atoms combine to form molecular orbitals that are filled with the molecule's electrons. The electron configuration depicted in these diagrams provides insights into molecular stability, magnetic properties, and potential reactivity.

As an example, consider the molecular orbital diagram for diatomic oxygen (O_2). Each oxygen atom contributes its atomic orbitals, leading to the formation of several σ and π bonding and antibonding molecular orbitals. Oxygen's ground-state electron configuration indicates that it has parallel-unpaired electrons in the π^* orbitals, accounting for its paramagnetic property—a behavior that valence bond theory cannot easily explain.

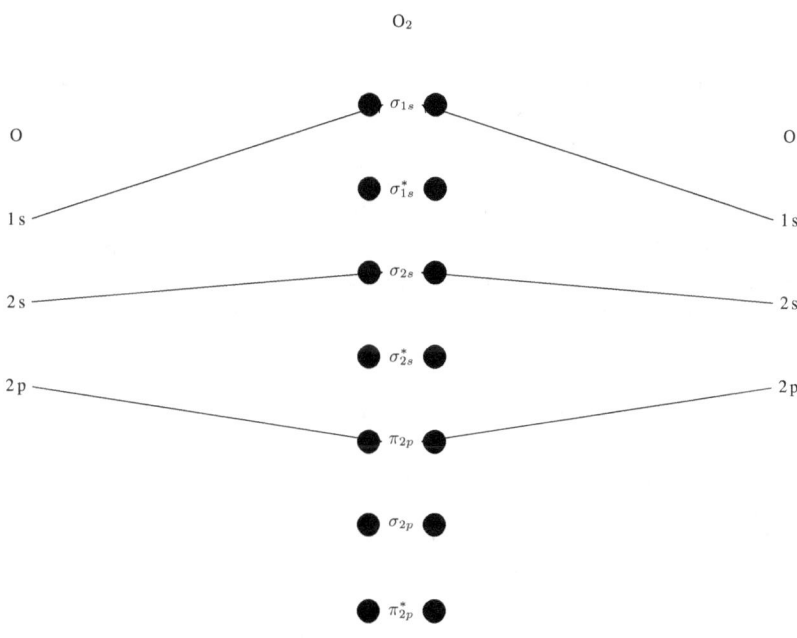

Ground state electronic configuration

- **Bond Order and Stability**

An essential concept derived from MO theory is the bond order, which provides a quantitative measure of bond strength and stability. Bond order is calculated as one-half the difference between the number of electrons in bonding molecular orbitals and those in antibonding molecular orbitals:

$$\text{Bond Order} = \frac{1}{2}\left(\text{Number of bonding electrons} - \text{Number of antibonding electrons}\right)$$

A positive bond order indicates a stable molecule, with a higher bond order implying a stronger bond. For example, in O_2, the bond order is 2, consistent with the presence of a double bond between the oxygen atoms.

- **Advantages and Applications of Molecular Orbital Theory**

MO theory provides numerous advantages over other bonding theories. It readily explains phenomena such as paramagnetism, as seen in oxygen, and the delocalization of electrons, exemplified by conjugated systems like benzene. Moreover, MO theory is indispensable in describing electronic transitions in spectroscopy, aiding in the interpretation of various spectroscopic methods, including UV-visible and infrared spectroscopy.

In quantum chemistry, MO theory is foundational in computational chemistry methods such as Hartree-Fock and Density Functional Theory (DFT), which rely on approximations and expansions of molecular orbitals to predict molecular properties and reactivity.

- **Complex Systems and Frontier Orbitals**

For more complex molecular systems, especially those involving multiple atoms and delocalized electrons, molecular orbital theory is particularly useful in describing frontier molecular orbitals—the highest occupied molecular orbital (HOMO) and the lowest unoccupied molecular orbital (LUMO). These frontier orbitals are crucial in determining the chemical reactivity and properties of molecules, as they participate actively in chemical reactions, especially in processes like electron transfer and catalysis.

The concept of frontier orbitals is integral to the understanding of photochemical reactions, where electronic excitation involves promotion of electrons from the HOMO to the LUMO, influencing reactivity patterns in processes such as polymerization and photosynthesis.

- **Visual Representation and Computational Approaches**

The advent of computer technology and graphical visualization techniques has transformed the practical application of molecular orbital theory. With software tools, chemists can visualize molecular orbitals, interpret their shapes, energies, and interactions, and obtain computationally derived insights into reacting systems and materials.

These computational approaches complement traditional experimental chemistry, allowing for predictions of reaction outcomes, property evaluation in drug design, catalyst development, and material science.

Molecular orbital theory provides a robust and versatile framework that extends beyond the capabilities of valence bond theory. It enables a profound understanding of electronic structure and molecular properties, facilitating both theoretical explorations and practical applications in various chemical disciplines. Through molecular orbital theory, chemists elucidate the underlying principles governing molecular behavior, advancing the field of chemistry and material science.

3.4 Hybridization and Molecular Geometry

Hybridization and molecular geometry are fundamental concepts in molecular chemistry, providing a framework for understanding the three-dimensional arrangement of atoms in molecules and their associated properties. Hybridization explains the formation and nature of covalent bonds, while molecular geometry describes the spatial configuration of these bonds, determining the overall shape and symmetry of the molecule.

Concept of Hybridization

Hybridization is a theoretical model that involves the mixing of atomic orbitals to form new hybrid orbitals. These hybrid orbitals have intermediate properties and energies compared to the original atomic orbitals, facilitating effective overlap and bond formation. The concept was introduced to rationalize molecular shapes that could not be adequately explained by simple atomic orbital interactions.

Hybridization considers the promotion of paired electrons in filled orbitals into higher energy, unfilled orbitals, enabling the formation of bonds that align with experimentally observed geometries. For example, carbon, with an electronic configuration of [He] $2s^2 2p^2$, can undergo hybridization to provide sp^3 orbitals, allowing it to form four equivalent bonds, as observed in methane (CH_4).

Types of Hybridization and Corresponding Geometries

- *sp Hybridization:* Involves the mixing of one s and one p orbital, resulting in two linearly oriented sp hybrid orbitals, separated by an angle of 180°. This hybridization is typical of molecules like acetylene (C_2H_2), where two sp hybrid orbitals form sigma bonds while the remaining p orbitals participate in pi bonds.

- *sp^2 Hybridization:* Results from the combination of one s and two p orbitals, generating three sp^2 hybrid orbitals arranged in a trigonal planar geometry with 120° interbond angles. This hybridization is characteristic of ethylene (C_2H_4), where the carbon atoms form sigma bonds using sp^2 orbitals and a pi bond via the unhybridized p orbitals.

- *sp^3 Hybridization:* Involves one s and three p orbitals, producing four equivalent sp^3 hybrid orbitals arranged tetrahedrally at 109.5°. This is seen in methane (CH_4), where each carbon atom forms single bonds with hydrogen atoms.

- *sp^3d Hybridization:* Combines one s, three p, and one d orbital, resulting in five sp^3d hybrid orbitals arranged in a trigonal bipyramidal geometry, with bond angles of 90° and 120°. Molecules such as phosphorus pentachloride (PCl_5) exhibit this type of hybridization.

- *sp^3d^2 Hybridization:* Consists of one s, three p, and two d orbitals, forming six sp^3d^2 hybrid orbitals in an octahedral geometry with 90° interbond angles, typical of sulfur hexafluoride (SF_6).

Understanding Molecular Geometry

Molecular geometry refers to the spatial arrangement of atoms in a molecule, critically influencing its physical and chemical properties, such as reactivity, polarity, phase of matter, color, magnetism, and biological activity. The shape of a molecule is determined by the number of electron domains around the central atom, considering both bonded atoms and lone pairs, as described by the VSEPR (Valence Shell Electron Pair Repulsion) theory.

3.4. HYBRIDIZATION AND MOLECULAR GEOMETRY

VSEPR theory asserts that electron domains, whether bonding or lone pairs, repel each other and arrange themselves to minimize this repulsion, resulting in specific geometric configurations. Here are some common geometric arrangements derived from VSEPR considerations:

- *Linear Geometry:* Characterized by two electron domains, leading to a linear arrangement with 180° between bonded atoms, as in carbon dioxide (CO_2).

- *Trigonal Planar Geometry:* Associated with three electron domains, creating a flat triangular shape with 120° angles, observed in boron trifluoride (BF_3).

- *Tetrahedral Geometry:* Involves four electron domains, producing a three-dimensional tetrahedron with 109.5° angles, exemplified by methane.

- *Trigonal Bipyramidal Geometry:* Encompasses five electron domains, forming two pyramids with a common triangular base, resulting in 90° and 120° angles. This is characteristic of phosphorus pentachloride (PCl_5).

- *Octahedral Geometry:* Consists of six electron domains, creating an octahedron with 90° angles as seen in sulfur hexafluoride.

- *Bent or Angular Geometry:* Occurs when there are lone pairs on the central atom, resulting in a bent shape. For instance, water (H_2O) with two hydrogen atoms and two lone pairs adopts a bent shape due to lone pair repulsion.

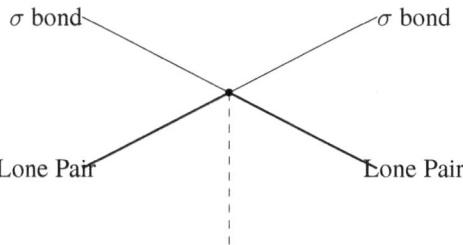

Hybridization in Complex and Multicenter Systems

For complex systems involving multicenter bonds or delocalized electronic structures, hybridization provides a basis for understanding interactions that are beyond simple sigma (σ) and pi (π) bonding. This is especially relevant in coordination chemistry and organic compounds featuring conjugated systems.

- *Conjugated Systems:* Delocalization of electrons in conjugated systems such as benzene (C_6H_6) can be explained by the resonance between different hybridized configurations. The sp^2 hybridization of each carbon atom in benzene allows for a planar arrangement with delocalized pi bonds above and below the molecular plane.

- *Coordination Complexes:* In complexes like $[Fe(CN)_6]^{4-}$, the iron ion undergoes sp^3d^2 hybridization, allowing formation of six coordinated covalent bonds with cyanide (CN^-) ligands in an octahedral geometry. Hybridization aids in visualizing ligand attachment and electronic distribution within metal complexes.

Impact of Molecular Geometry on Properties

The geometry of a molecule has profound implications for its polarity, intermolecular interactions, and reactivity. Polar molecules, which result from asymmetric charge distribution, exhibit dipole moments influencing solubility and intermolecular forces.

- *Polarity:* Molecules like water (H_2O) are polar due to the bent geometry and electronegative oxygen atom, leading to hydrogen bonding critical in biological and physical processes. In contrast, linear carbon dioxide is nonpolar despite containing polar bonds due to symmetrical geometry canceling dipole moments.

- *Reactivity:* The accessibility and exposure of reactive sites in molecules are influenced by geometry. For instance, tetrahedral molecules such as tetraethyllead ($(C_2H_5)_4Pb$) feature steric hindrance reducing reactivity. Conversely, linear molecules with exposed reactive sites tend to be more reactive.

- *Intermolecular Forces and Phase Behavior:* Molecular geometry affects the types and strengths of intermolecular forces, including dipole-dipole interactions, London dispersion forces, and hydrogen bonding.

Polyatomic gaseous molecules, such as NH_3, demonstrate higher boiling points due to polarity and hydrogen bonding induced by molecular geometry.

Analyzing Molecular Geometry and Hybridization with Modern Techniques

Understanding hybridization and geometry in complex molecules benefits from modern analytical techniques and computational modeling. Techniques such as X-ray crystallography, nuclear magnetic resonance (NMR) spectroscopy, and infrared (IR) spectroscopy provide precise data on molecular structures, while computational methods including density functional theory (DFT) simulate electronic properties and geometrical configurations accurately.

Utilizing these tools, chemists can explore and rationalize variations in hybridization and geometry that affect reactivity patterns, environmental impacts, and potential applications in new material development, pharmaceuticals, and catalysis.

Understanding these concepts enables scientists to unravel the complexities of chemistry at a fundamental level, influencing diverse fields such as material science, biochemistry, and nanotechnology.

3.5 Intermolecular Forces

Intermolecular forces are the electrostatic forces of attraction and repulsion between molecules or between molecules and ions. These forces are distinguishable from intramolecular forces, which are the forces within a molecule that hold chemical bonds together. Understanding intermolecular forces is essential for explaining the physical properties of substances, such as boiling and melting points, solubility, viscosity, and surface tension.

Intermolecular forces are generally weaker than covalent or ionic bonds, but they are integral to the behavior and interaction of molecules in various states of matter. The primary types of intermolecular forces include London dispersion forces, dipole-dipole interactions, hydrogen bonding, and ion-dipole interactions.

London Dispersion Forces

London dispersion forces, also known as van der Waals forces, are the weakest of the intermolecular forces and arise due to transient dipoles that occur as a result of temporary fluctuations in the electron density distribution within a molecule. These temporary dipoles can induce similar dipoles in neighboring molecules, resulting in an attractive interaction.

Dispersion forces are present in all molecules, whether polar or nonpolar, but are particularly significant in the noble gases and nonpolar molecules such as N_2 or O_2. The strength of dispersion forces correlates with the polarizability of a molecule, which refers to the ease with which the electron distribution can be distorted. Larger and more massive atoms or molecules with more electrons exhibit greater polarizability and therefore stronger dispersion forces. For example, the boiling point of noble gases increases with increasing atomic number from helium (He) to xenon (Xe), reflecting stronger dispersion forces.

Due to their ubiquitous nature, London dispersion forces play a critical role in the condensation of gases and the overall stability of molecular systems in gaseous and liquid phases.

Dipole-Dipole Interactions

Dipole-dipole interactions occur between polar molecules, where the partial positive charge on one molecule is attracted to the partial negative charge on another. These interactions depend on the dipole moment, a measure of molecular polarity, and are typically stronger than London dispersion forces but weaker than hydrogen bonds.

Molecules with significant dipole moments, such as hydrogen chloride (HCl) or sulfur dioxide (SO_2), exhibit more substantial dipole-dipole interactions that influence their boiling and melting points. The orientation of polar molecules in an electric field or under the influence of other dipoles can lead to structured arrangements and specific physical properties.

In liquid states, dipole-dipole interactions contribute to an organized molecular structure, affecting viscosity and surface tension. Such interactions also play a role in the solubility of polar substances in polar solvents, following the adage "like dissolves like."

Hydrogen Bonding

Hydrogen bonding is a special case of dipole-dipole interaction, occurring between a hydrogen atom covalently bonded to a highly electronegative atom (commonly nitrogen, oxygen, or fluorine) and the lone pair of electrons on another electronegative atom. This type of bonding is notably stronger than typ-

3.5. INTERMOLECULAR FORCES

ical dipole-dipole interactions, significantly affecting the properties of compounds.

Water (H_2O), with its ability to form up to four hydrogen bonds, is a classic example where hydrogen bonding leads to high boiling and melting points relative to its molecular weight. These bonds are responsible for many of water's unique properties, such as its high specific heat, surface tension, and its solid-liquid phase density anomaly, where ice is less dense than liquid water.

Hydrogen bonding also plays a vital role in biological systems, contributing to the structure and function of nucleic acids like DNA and proteins. The double helix structure of DNA is stabilized by hydrogen bonds between the complementary base pairs, while protein secondary structures, including alpha helices and beta sheets, rely on hydrogen bonding for stability.

Ion-Dipole Interactions

Ion-dipole interactions occur between ionic species and polar molecules. These interactions are crucial in solutions where ions interact with solvent molecules. The strength of ion-dipole forces depends on the charge and size of the ion and the dipole moment of the polar molecule.

In aqueous solutions, ion-dipole interactions explain much of the dissolution process of ionic salts such as sodium chloride (NaCl). The chloride and sodium ions interact with the polar water molecules, with the hydrogen atoms facing chloride anions and oxygen atoms facing sodium cations, effectively stabilizing them in solution.

Ion-dipole forces are stronger than hydrogen bonds due to their electrostatic nature involving full ionic charges and partially charged dipoles, thus significantly influencing solubility and the ability of solvents to dissolve ionic compounds.

Interplay and Relative Strength of Intermolecular Forces

The interplay between various intermolecular forces determines the macroscopic properties of substances. While the strength of these forces generally follows: London dispersion < dipole-dipole < hydrogen bonding < ion-dipole, it is important to note that a substance may experience multiple types of intermolecular interactions simultaneously.

These forces dictate numerous physico-chemical properties, such as:

- *Boiling and Melting Points:* Stronger intermolecular forces imply higher boiling and melting points. For example, hydrogen fluoride (HF) boils at much higher temperatures than fluorine (F_2) due to hydrogen bonding, despite similar molecular weights.

- *Viscosity and Surface Tension:* Substances with strong intermolecular forces like hydrogen bonding (e.g., glycerol) exhibit higher viscosity and surface tension, affecting their flow and spreading behavior.

- *Solubility:* Solubility patterns are strongly influenced by intermolecular interactions, with polar solutes typically dissolving well in polar solvents due to dipole-dipole and hydrogen bonding, whereas non-polar solutes dissolve better in non-polar solvents through dispersion forces.

- *Phase Transitions and Graphical Representation:* The determination of phase diagrams often considers the balance of various intermolecular forces, predicting states under different temperature and pressure conditions.

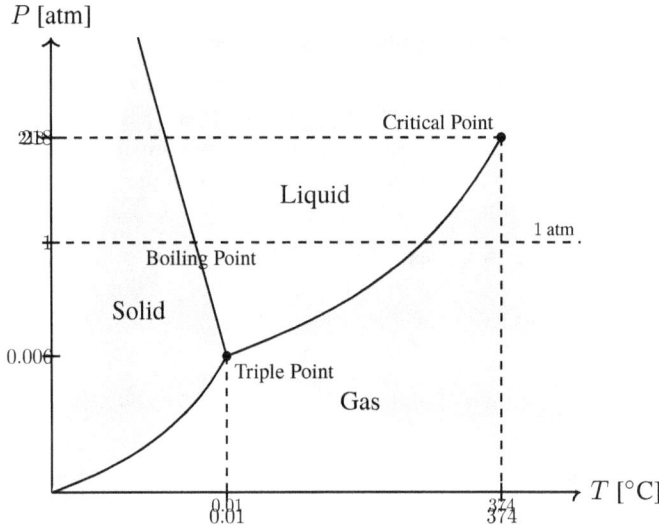

Technological and Biological Implications

Understanding intermolecular forces enables advancements across various fields, including material science, pharmacology, and environmental science. In the material sciences, designing polymers and composites with targeted adhesion, flexibility, or thermal properties relies on tailoring intermolecular interactions.

In medicine, intermolecular forces drive drug-receptor interactions that underlie therapeutic efficacy. Drug molecules often form temporary bonds with biological macromolecules, guided by hydrogen bonds and hydrophobic interactions, influencing pharmacodynamics and pharmacokinetics.

Environmental phenomena such as the surface tension of water affecting plant capillary action, aerosols and droplets in cloud formation, and pollution diffusion in the atmosphere rely on the interplay of these forces. New green technologies aim to mimic biological systems utilizing intermolecular forces for sustainable energy and resource use.

Intermolecular forces represent an integral component of chemical understanding, connecting atomic-level interactions to observable macroscopic behaviors. A grasp of these forces deepens appreciation for the complex balance present in natural and engineered systems, providing a foundation for innovation and harnessing chemistry toward societal needs.

3.6 Polarity and Its Effects on Properties

Polarity is a fundamental concept in chemistry that describes the distribution of electric charge across a molecule. This distribution results in molecules having a positive and a negative pole, much like a magnet, and is primarily determined by the molecule's geometry and the electronegativity differences between its atoms. The concept of polarity is essential for understanding a wide range of chemical behaviors and properties, from solubility to boiling and melting points, and from reactivity to intermolecular interactions.

Understanding Molecular Polarity

Polarity arises when there is an uneven distribution of electron densities in a molecule. This unevenness is typically due to differences in electronegativity, which is the ability of an atom within a molecule to attract electrons toward itself. When two atoms of differing electronegativity form a covalent bond, the shared electrons are displaced towards the more electronegative atom, creating a dipole moment, where one end of the bond or molecule is more negative (due to higher electron density) and the other end is more positive.

Determining Molecular Polarity

The determination of molecular polarity involves two primary considerations: the polarity of individual bonds and the geometry of the molecule.

- *Bond Polarity:* The polarity of each bond within a molecule is evaluated based on the electronegativity difference between the bonded atoms. A bond is considered polar if the electronegativity difference is significant, typically greater than 0.5 on the Pauling scale. For instance, in hydrogen chloride (HCl), chlorine is more electronegative than hydrogen, making the bond polar.

- *Molecular Geometry:* Regardless of individual bond polarities, the overall shape of the molecule determines the net dipole moment. Molecular geometry describes the three-dimensional arrangement of atoms in a molecule. If the geometry is symmetric, the dipole moments of individual bonds may cancel each other out, resulting in a nonpolar molecule. In contrast, asymmetric molecules often have a net dipole moment. Carbon dioxide (CO_2) is linear and symmetrical, making it nonpolar, while water (H_2O), with its bent shape, is polar.

Visualizing Dipole Moments

The dipole moment (μ) is a vector quantity that represents the separation of charge within a molecule and is mathematically expressed as:

$$\mu = \delta \cdot d$$

where δ is the magnitude of the partial charges, and d is the distance between the charges. Dipole moments are measured in debye units (D), where 1 D is equivalent to 3.34×10^{-30} C $\cdot$ m.

3.6. POLARITY AND ITS EFFECTS ON PROPERTIES

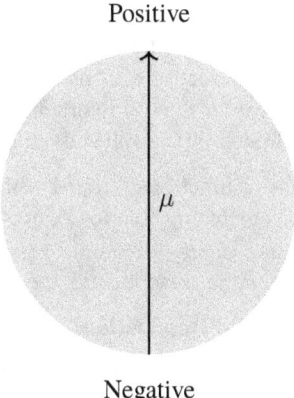

Positive

Negative

Effects of Polarity on Physical Properties

- *Boiling and Melting Points:* Polar molecules typically exhibit higher boiling and melting points than nonpolar molecules of similar size due to stronger intermolecular forces, such as dipole-dipole interactions and hydrogen bonds. For example, hydrogen sulfide (H_2S) boils at a lower temperature than water (H_2O), despite being heavier, because water molecules form extensive hydrogen bonds.

- *Solubility:* Polarity significantly affects a substance's solubility, governed by the principle "like dissolves like." Polar substances tend to dissolve well in polar solvents (e.g., ionic compounds and polar organic molecules in water), while nonpolar substances dissolve better in nonpolar solvents (e.g., oils and fats in hexane).

- *Surface Tension and Capillarity:* Water's high surface tension and capillary action are due to its polarity, which promotes hydrogen bonding. These properties are crucial for biological processes like nutrient transport in plants and insects walking on water surfaces.

- *Vapor Pressure:* Polar substances often have lower vapor pressures compared to nonpolar substances, primarily due to strong intermolecular forces that hinder the escape of molecules into the vapor phase.

Chemical Reactivity and Catalysis

Polarity also plays a critical role in determining molecular reactivity and mechanisms in chemical reactions. Reaction intermediates, transition states, and catalysts often rely on polar interactions. Consider acid-base reactions, where the electrophilic nature of protons (H^+) interacts with nucleophiles (OH^-) in aqueous solutions, facilitated by the solvent's polarity.

In catalysis, polar domains or functional groups within catalysts can stabilize polarized transition states, enhancing reaction rates. The design of catalysts in processes like hydrogenation (using platinum or palladium) takes advantage of polar interactions between the substrate and the catalyst surface.

Biological and Environmental Implications

The influence of polarity extends into biology and environment, affecting molecular recognition, transport processes, and pollutant dispersion.

- *Molecular Recognition:* Polarity affects the selective binding between enzymes and substrates, or between antibodies and antigens, through non-covalent interactions like hydrogen bonds and ionic interactions.

- *Transport and Membrane Permeability:* Cell membranes possess a polar and nonpolar nature, facilitating selective permeability. Polar molecules such as water and ions require transport proteins to traverse the lipid bilayer, whereas nonpolar compounds diffuse directly.

- *Environmental Fate and Transport of Chemicals:* The polarity of environmental pollutants determines their transport and distribution. Nonpolar hydrocarbons often accumulate in water-insoluble phases, while polar compounds distribute more readily in aquatic environments, influencing contamination and remediation strategies.

Analyzing Polarity and Applications in Technology

Modern analytical techniques and computational methods aid in characterizing molecular polarity. Techniques like nuclear magnetic resonance (NMR) spectroscopy, infrared (IR) spectroscopy, and X-ray crystallography provide insights into molecular polarity and structures. Computational chemists use molecular simulations and quantum mechanical calculations to predict dipole moments and polar interactions, assisting in the design of pharmaceuticals and novel materials.

- *Pharmaceuticals:* Drug design exploits molecular polarity for optimizing drug solubility, absorption, and target binding. Polar functional

3.7. RESONANCE AND DELOCALIZED ELECTRONS

groups improve pharmacokinetic profiles, increasing drug efficacy and reducing side effects.

- *Materials Science:* Polar properties are exploited in the development of conductive polymers, liquid crystals, and piezoelectric materials, where polarity dictates electronic properties, phase behaviors, and mechanical responses.

- *Nanotechnology and Coatings:* Advances in nanotechnology utilize polar functionalities for surface modification, improving adhesion, compatibility, and functionality in nanocomposites, coatings, and lithography processes.

The intricate relationship between polarity and the physicochemical properties of molecules emphasizes its fundamental importance in chemistry and related disciplines. Appreciating polarity advances the understanding of molecular behavior in diverse contexts, facilitating scientific discoveries and technological innovations.

3.7 Resonance and Delocalized Electrons

Resonance and the concept of delocalized electrons are pivotal in understanding the chemical behavior and stability of certain molecules. These concepts provide insight into the electronic structure of molecules where single Lewis structures fail to adequately describe the electron distribution. Resonance involves presenting a molecule with multiple plausible structures, while electron delocalization refers to the spread of electrons over several atoms, contributing to the overall stability of the molecule.

Understanding Resonance

Resonance arises when a molecule can be represented by two or more valid Lewis structures that differ only in the placement of electrons, not in the arrangement of atoms. These alternative structures, known as resonance contributors or resonance forms, collectively describe the real electron distribution within the molecule better than any single structure can. While individual resonance forms are hypothetical, the resonance hybrid, a weighted average of these contributors, represents the actual electron distribution more accurately.

Key features of resonance include:

- The positions of atoms remain fixed across different resonance forms.
- Only the placement of electrons, particularly pi electrons and lone pairs, changes between forms.
- Resonance structures must obey the rules of covalent bonding, including accommodating the octet rule where applicable.
- The resonance hybrid is generally more stable than any single resonance form, due to electron delocalization minimizing potential energy.

Common Examples and Their Resonance Structures

- *Benzene (C_6H_6):* One famous example of resonance is benzene, where the pi electrons are delocalized across the six carbon atoms of the ring. The two primary resonance forms feature alternating single and double bonds, but the actual structure is a resonance hybrid with equal bond lengths, often depicted as a hexagon with an inscribed circle representing delocalized electrons.
- *Acetate Ion (CH_3COO^-):* The acetate ion exhibits resonance as the negative charge and pi bond are shared between the two oxygen atoms. Both oxygen atoms can alternately hold the double bond and the negative charge, offering two main resonance contributors.
- *Nitrate Ion (NO_3^-):* In nitrate, the three resonance forms show the negative charge and pi bonding distributed among the three oxygen atoms. This delocalization helps explain the equivalent bond lengths observed experimentally.

$$\begin{aligned}
\text{Benzene:} &\quad \text{Kekulé structures} \rightleftharpoons \text{Resonance Hybrid} \\
\text{Acetate:} &\quad \text{Resonance structure A} \rightleftharpoons \text{Resonance structure B} \\
\text{Nitrate:} &\quad \text{Structure X} \rightleftharpoons \text{Structure Y} \rightleftharpoons \text{Structure Z}
\end{aligned}$$

Energetic Implications of Resonance

The stabilization offered by resonance is primarily due to the delocalization of electrons, which can lower the potential energy of the molecule. This resonance stabilization is noticeable in comparison to hypothetical non-resonance

3.7. RESONANCE AND DELOCALIZED ELECTRONS

scenarios and is often quantified in terms of resonance energy, the difference in energy between the actual structure and the most stable predicted structure without resonance.

For example, benzene exhibits significant resonance stabilization, contributing to its unique chemical properties, such as resistance to addition reactions typical for alkenes. This stabilization is evidenced by its high resonance energy, supporting the idea that delocalized electrons contribute to the molecule's lower energy state.

Rules and Limitations in Drawing Resonance Structures

While resonance structures are a powerful tool for describing electron distribution, their construction must adhere to certain guidelines:

- Only electrons, not atoms, may be moved between resonance forms.
- The overall charge of the molecule must remain constant across all forms.
- Each resonance structure must be a valid Lewis structure.
- Structures with complete octets and minimal charge separation are generally favored.

While resonance provides critical insight, it is inherently limited to pi bonds and lone pairs. Certain significance is given to structures where octets are complete and that exhibit minimal charge separation, aligning with actual molecular properties more closely.

Delocalized Electrons and Conjugation

Delocalization of electrons refers to the phenomenon where electrons, particularly pi electrons, are shared across multiple atoms, leading to a more distributed electron cloud. This phenomenon is critical in conjugated systems, characterized by alternating single and double bonds, which allow pi electrons to be delocalized over the conjugated pathway.

Conjugated systems are prevalent in organic chemistry and are responsible for various chemical and physical properties:

- *Optical Properties:* Conjugated systems exhibit distinctive optical properties, including absorption of UV and visible light. This property is exploited in dyes and pigments, where conjugated chromophores absorb specific wavelengths, imparting color.

- *Stability:* Delocalization confers additional stability to conjugated molecules, evident in lower reactivity of conjugated dienes compared to isolated counterparts.

- *Reactivity:* Conjugated molecules often participate in unique reactions, such as Diels-Alder cycloaddition or electrophilic aromatic substitution, where delocalization stabilizes intermediate species.

Conjugation Pathway

Applications and Examples in Advanced Materials

The principles of resonance and electron delocalization underpin the design of novel materials with specific electronic and structural properties. Conjugated polymers, often utilized in organic electronics, leverage delocalization for enhanced electrical conductivity. These polymers, featuring extensive conjugated systems, offer flexibility and processability, opening avenues in solar cells, light-emitting diodes, and field-effect transistors.

Conductive polymers such as polyaniline and polyacetylene exhibit remarkable conductivity due to delocalized electrons that can move freely along the conjugated chains. These materials are integral to developing thin, flexible electronic devices, biosensors, and smart textiles.

Biological Significance and Structural Functionality

In biology, electron delocalization is crucial for the function and stability of macromolecules such as heme, chlorophyll, and nucleic acids. The highly conjugated porphyrin ring system in heme allows for electron transfer events essential to oxidative phosphorylation. Similarly, chlorophyll's conjugated system facilitates the capture of light energy for photosynthesis.

The delocalization of electrons in nucleic acids contributes to the stabilization of DNA and RNA structures through aromatic stacking interactions, critical for maintaining genetic information integrity and enabling the structural transitions involved in replication and transcription.

3.7. RESONANCE AND DELOCALIZED ELECTRONS

Challenges and Interpretations in Computational Chemistry

Modeling resonance and electron delocalization presents challenges due to the need for accurate descriptions of pi-electron distributions. Computational methods facilitate the interpretation of these effects by employing molecular orbital theory to predict electronic structure and energy contributions.

Difference density analysis, or quantum chemical topological approaches, allow for detailed exploration of electron delocalization. These methods aid in visualizing electron distribution and potential energy surfaces, deepening understanding of structural and electronic properties.

To summarize, resonance and delocalized electrons are indispensable constructs in modern chemistry, offering explanations for a myriad of chemical phenomena. Their roles in structural and electronic properties extend across chemical disciplines, underpinning advancements in materials science, biochemistry, and theoretical chemistry. Embracing these concepts enables chemists to harness electron mobility and stability, fostering innovations that reverberate throughout scientific and industrial domains.

Chapter 4

Thermodynamics and Equilibrium

Thermodynamics formalizes the study of energy transformations and the directionality of processes within chemical systems, laying the foundation for understanding equilibria in physical and chemical contexts. This chapter traverses the laws of thermodynamics, elucidating the principles that dictate how energy transfer influences system states and potentials. Concepts such as enthalpy, entropy, and free energy are central to quantifying the spontaneity and feasibility of reactions and transitions, ultimately guiding the comprehension of how equilibrium is achieved and maintained. Through the analytical lens of phase equilibria and Le Chatelier's principle, the interdependencies among temperature, pressure, and concentration are unraveled, revealing the delicate balance that governs systemic stability.

4.1 Fundamental Concepts of Thermodynamics

Thermodynamics is the branch of physics that deals with the relationships between heat and other forms of energy. In the context of thermodynamics, a profound understanding of systems, states, and processes is essential. Thermodynamics serves as a foundational pillar in chemistry, physics, and engineering, providing critical insights into how energy is transferred and transformed within different types of systems. This section delves into these essential concepts with meticulous detail, ensuring clarity and comprehension.

A *thermodynamic system* is defined as the specific portion of the universe that is being studied. This system can be as simple as a single chemical reaction or as complex as an entire power plant. There are three primary types of thermodynamic systems: isolated, closed, and open systems.

An *isolated system* exchanges neither energy nor matter with its surroundings. The coffee in a perfectly insulated thermos can be considered an isolated system over a short time frame, even though a perfectly isolated physical container does not exist in practice.

An *closed system* can exchange energy but not matter with the surroundings. An example of a closed system is a sealed steam radiator: while water vapor may condense and cool, neither water nor air enters or leaves the system.

An *open system* can exchange both energy and matter with its surroundings. Biological organisms exemplify open systems as they intake food (matter) and release heat and waste products (energy and matter).

The boundaries that differentiate these systems can be either real or imaginary and may be flexible or rigid, stationary or moving. Defining the type of system being analyzed is crucial for applying the principles of thermodynamics appropriately.

The *state* of a thermodynamic system is its condition at a specific point in time, defined by state variables such as pressure, volume, temperature, and composition. Each state is an ensemble of properties that offer a complete description of the system at a moment. In thermodynamics, systems are assumed to be in a state of equilibrium unless a change is applied.

When we speak of *processes* in thermodynamic terms, we refer to the transformation of one state into another state. Processes involve changes in one or more state variables and can be delineated as isothermal, isobaric, isochoric, or adiabatic, based on the constant that remains unchanged during the process.

An *isothermal process* occurs at a constant temperature. During this process, heat is either supplied to or removed from the system to maintain the temperature as constant. A common example is the phase transition between liquid and gas, such as boiling water at atmospheric pressure.

An *isobaric process* takes place at constant pressure. If a gas in a cylinder is

4.1. FUNDAMENTAL CONCEPTS OF THERMODYNAMICS

heated such that the piston adjusts to keep the pressure constant, then the process is isobaric. These processes are especially significant in understanding the working principles of heat engines and refrigerators.

An *isochoric process* occurs at a constant volume. No work is done since the volume does not change, making such processes particularly significant in situations where volumetric expansion or compression doesn't occur.

An *adiabatic process* is carried out without any heat exchange with the environment. Such processes happen over an insulated boundary, making them ideal models for fast processes where no heat transfer has time to occur, such as the rapid compression in an internal combustion engine.

The *principle of conservation of energy* is fundamental to thermodynamics and encompasses the first law, which states that the change in the internal energy of a system is equal to the heat added to the system minus the work done by the system on its surroundings. This can be formally expressed as:

$$\Delta U = Q - W$$

where ΔU represents the change in internal energy, Q is the heat added to the system, and W is the work done by the system. This equation underscores the notion that energy cannot be created or destroyed—only transformed.

Next, *state functions* and *path functions* are intrinsic to understanding thermodynamic processes. State functions, such as internal energy, enthalpy, and entropy, depend solely on the current state of the system, not on the path by which the system arrived at that state. On the contrary, path functions like work and heat depend on the specific transitions a system experiences between two states.

Insights into state functions were propelled by the understanding of *enthalpy* (H), defined as:

$$H = U + PV$$

where U is the internal energy, P the pressure, and V the volume. Enthalpy is crucial in processes occurring at constant pressure, providing a measure of the total energy, including contributions from both the internal energy and the energy required to make space for the system's surroundings.

Enthalpy changes (ΔH) provide valuable insights into chemical reactions and phase changes. An exothermic reaction results in a negative ΔH due to the release of heat to the surroundings, whereas an endothermic reaction exhibits a positive ΔH as it absorbs heat.

The concept of *PV-work* and *expansion work* is pivotal to processes where the system's volume changes, which alters the system's capacity to perform work on its surroundings. In processes where gases expand or are compressed, work is done via these pathways. The volume changes can be calculated using:

$$W = -\int_{V_i}^{V_f} P\,dV$$

This integral calculates the work done by a system as it expands from an initial volume V_i to a final volume V_f, with P being the pressure at each incremental volume change.

Lastly, the conceptualization of *thermodynamic equilibrium* establishes that a system is in equilibrium if the state functions remain constant over time, and there are no net flows of matter or energy. Equilibrium is achieved when all processes occur reversibly and simultaneously in infinite small steps—an idealization used in theoretical models.

Through this engagement with fundamental concepts, the groundwork is laid for further explorations of the laws and principles that govern energy exchanges within systems, setting up the analytical tools required to delve into more advanced thermodynamic analyses.

4.2 The Laws of Thermodynamics

The laws of thermodynamics form a framework for understanding the energy transformations and equilibrium properties within physical systems. They provide a coherent set of principles that elucidate how energy is converted, transferred, and conserved, establishing the underpinnings for disciplines ranging from physics to chemistry and engineering. These laws—covering the zeroth, first, second, and third laws—serve not only to describe empirical observations but also to constrain the nature of physical reality itself.

The *zeroth law of thermodynamics* provides the foundational definition of temperature and thermal equilibrium. It states that if two systems, say A and B, are each in thermal equilibrium with a third system C, then A and B must be

4.2. THE LAWS OF THERMODYNAMICS

in thermal equilibrium with each other:

$$A \equiv C \quad \text{and} \quad B \equiv C \implies A \equiv B$$

The zeroth law allows for the establishment of temperature as a meaningful property that can be measured, and it implies the existence of a temperature scale. It also introduces the concept of thermodynamic surfaces or manifolds, where every point corresponds to a state of thermal equilibrium.

The *first law of thermodynamics*, also known as the law of energy conservation, posits that energy can neither be created nor destroyed. It acknowledges that the total energy of an isolated system remains constant, although energy can change forms, for instance, from chemical energy to kinetic energy. This law can be expressed via the equation:

$$\Delta U = Q - W$$

where ΔU represents the change in internal energy, Q is the heat added to the system, and W is the work done by the system on its surroundings. The first law asserts the equivalence of work and heat as forms of energy transfer and introduces the concept of internal energy as a state function.

Consider as an example a gas contained in a piston. When the gas absorbs heat (such as through combustion in an engine), it expands, performing work on the piston. Alternatively, when compressed, work is done on the gas, increasing its internal energy and temperature. The first law provides the exact balance between the heat supplied to the gas and the work done by it during this transformation.

The *second law of thermodynamics* deals with the directionality of processes and the concept of entropy as a measure of disorder or randomness. It states that for any spontaneous or natural process, the entropy of an isolated system will always increase. In more specific terms, the second law declares that energy conversions are not 100% efficient and that there is always an increase in entropy:

$$\Delta S \geq 0$$

where S is the entropy of the system. The equality holds for a reversible process, and the inequality becomes strict for irreversible processes. This law implies that spontaneous processes in the universe move toward thermodynamic equilibrium, characterized by maximum entropy.

A classic example of the second law in action is the mixing of two gases. When a barrier between two gases in a container is removed, the gases will spontaneously mix, leading to an increase in entropy and achieving an energetically uniform state. This unfailing tendency towards increased entropy suggests a direction for processes and determines the feasibility of energy transformations.

Additionally, the second law introduces the concept of *heat engines* and the *Carnot cycle*, which dictates that no heat engine operating between two heat reservoirs can be more efficient than a Carnot engine:

$$\eta = \frac{W}{Q_H} = 1 - \frac{T_C}{T_H}$$

where η is the efficiency, W is the work output, Q_H is the heat absorbed, and T_C and T_H are the temperatures of the cold and hot reservoirs, respectively.

The irreversibility implied by the second law of thermodynamics results in the generation of waste energy, often manifesting as heat, that cannot be completely converted back into useful work, embodying the principle of an intrinsic, unidirectional flow of time from less probable to more probable states.

The *third law of thermodynamics* ascribes an absolute frame to entropy. It establishes that as a system approaches absolute zero (0 K), the entropy of a perfectly crystalline substance approaches zero:

$$\lim_{T \to 0} S = 0$$

This law discerns the unattainability of absolute zero, implying constraints on cooling processes. The third law also provides a reference point for entropy, allowing for the calculation of absolute entropies for substances at any temperature above 0 K. Furthermore, it implies that as temperature decreases, so too does the capacity for energy dispersal within the system, leading to a determination that some entropy content becomes non-existent at 0 K.

To illustrate, consider the alignment of atoms within a crystalline lattice as a material is cooled. At temperatures near absolute zero, the vibration of atoms diminishes, eventually freezing in a perfect state of minimal entropy. Despite this approach to orderliness, reaching absolute zero is theoretically out of reach due to constraints related to the energy required to maintain continual cooling.

The laws of thermodynamics form a vast and consistent basis for understanding biological, chemical, and physical processes. They encompass concepts ranging from heat transfer mechanics and energy conservation to the natural trend towards disorder, laying the foundation for numerous practical applications.

These laws govern the efficiency of engines, the thermodynamic feasibility of reactions, and the constraints on energy extraction and utilization. Fundamentally, thermodynamics facilitates a comprehension of how the universe's energy is organized and guides advancements in technology and science through a robust framework spanning microscopic to macroscopic phenomena. The interrelation of these laws provides insight into how energy and matter behave, crafting a philosophy that underpins the study of nature across diverse scientific terrains.

4.3 Enthalpy and Heat Capacity

The foundational concepts of enthalpy and heat capacity are central to understanding heat transfer and energy transformations in thermodynamics, particularly in the context of chemical reactions and phase changes. These concepts enable us to quantify the energy absorbed or released during these processes, providing insights into reaction spontaneity and stability. A thorough comprehension of enthalpy and heat capacity also allows for the deduction of thermal properties of substances, playing a crucial role in both theoretical and practical applications within chemistry and physics.

Enthalpy, denoted as H, is a thermodynamic potential often used to account for heat flow in processes occurring at constant pressure. It is defined as:

$$H = U + PV$$

where U is the internal energy of the system, P is the pressure, and V is the volume. The change in enthalpy, ΔH, provides a measure of the heat absorbed or released in a process at constant pressure:

$$\Delta H = \Delta U + \Delta(PV)$$

Enthalpy is particularly useful in chemical reactions, where it allows chemists to discern whether a reaction is endothermic (absorbs heat) or exothermic (releases heat). Exothermic reactions, characterized by negative ΔH, result in the evolution of heat, exemplified by combustion reactions such as:

$$CH_4 + 2\,O_2 \longrightarrow CO_2 + 2\,H_2O + \Delta H$$

Conversely, endothermic reactions, which show positive ΔH, require heat absorption, as seen in the decomposition of calcium carbonate:

$$CaCO_3 \longrightarrow CaO + CO_2 - \Delta H$$

The *standard enthalpy change* ($\Delta H°$) refers to the change in enthalpy when a reaction occurs under standard conditions (1 atm pressure, 298.15 K). These standardized conditions allow for the consistency needed to tabulate and compare enthalpy changes for different reactions.

The enthalpy of reaction for any process is calculated using Hess's Law, which asserts that the total enthalpy change for a reaction is the same, irrespective of the path taken, relying on the property of enthalpy as a state function. Hess's Law can be used for reactions that need to be divided into multiple steps:

$$\Delta H_{total} = \sum \Delta H_{steps}$$

For example, if the direct conversion from reactants to products is not feasible, we might break the reaction down numerically and calculate ΔH for each step.

Heat capacity, symbolized as C, is an extensive property that describes the amount of heat required to change the temperature of a given substance by a unit degree. It is defined by the relation:

$$q = C\Delta T$$

where q is the heat absorbed or released, and ΔT is the temperature change. Heat capacity is critical in predicting how a substance responds to the input of energy in the form of heat.

There are two types of heat capacities to consider:

4.3. ENTHALPY AND HEAT CAPACITY

- *Specific heat capacity* (c) is the heat required to change the temperature of one gram of a substance by one degree Celsius or Kelvin. It is expressed as:

$$q = mc\Delta T$$

where m is the mass, illustrating its use in calculating the heat change for known masses.

Specific heat capacity varies between substances due to differing molecular structures and bonding energies. Water, for instance, has a high specific heat capacity (4.18 J · g^{-1} · K^{-1}), meaning it requires more energy to change its temperature than many other substances. This property makes water highly effective for thermal regulation in environments and living organisms.

- *Molar heat capacity* (C_m) describes the amount of heat needed to change the temperature of one mole of a substance by one degree Celsius or Kelvin and is given in J/mol·K:

$$q = nC_m\Delta T$$

where n is the number of moles. Molar heat capacity is similarly used to characterize substances at a molecular level, providing a deeper understanding of their energetic characteristics.

Under constant volume or pressure, heat capacity is referred to as C_v or C_p, respectively, and they are connected by the relation involving the gas constant R:

$$C_p = C_v + R$$

where R is the universal gas constant (8.314 J/mol · K). This relation holds more rigorously for ideal gases, establishing a basis for studying real gases and approximating their behaviors through similar relations.

The knowledge of heat capacities aids in the computation of energy requirements for heating and cooling processes, pivotal within engineering applications such as HVAC systems, refrigeration, and even in the design of combustion engines.

In chemical thermodynamics, both enthalpy and heat capacity are indispensable for probing reaction kinetics and mechanisms, interpreting calorimetric data, and optimizing industrial processes. For instance, in calorimetry—a method used to determine heat changes during a physical or chemical process—the calculated q based on measured temperature changes and known heat capacities provides direct insights into reaction energetics.

Furthermore, the heat capacity can be temperature-dependent, expressed as:

$$C = a + bT + cT^2 + \ldots$$

where a, b, and c are empirical coefficients. This polynomial expansivity allows fitting of heat capacity across a range of temperatures, enriching comprehension of how substances behave under diverse thermal conditions.

Exploring the *phase changes* further illustrates the significance of enthalpy and heat capacity. For example, when a substance transitions from solid to liquid or liquid to gas, the latent heat of fusion or vaporization is absorbed without temperature change, representing enthalpy changes during phase transitions.

Enthalpy and heat capacity extend into ecology and environmental science domains, aiding in assessing the thermal stability of ecosystems under climate perturbations. They are pivotal in modeling heat exchanges within the Earth's atmosphere and oceans, forecasting climate dynamics, and solving challenges presented by global warming.

Overall, enthalpy and heat capacity form the bedrock for understanding thermodynamic phenomena at both micro and macro levels, essential for disciplines such as physical chemistry, engineering, environmental science, and more, creating an unbroken continuum from fundamental science to applied technology. They characterize how energy transitions reshape the physical world, paving the way for innovative solutions across scientific and industrial fields.

4.4 Entropy and the Second Law

Entropy and the second law of thermodynamics are profound principles governing the directionality of natural processes and the evolution of energy systems. Entropy quantifies the potential disorder and randomness associated with a system, providing insight into its thermodynamic spontaneity and equilibrium. The second law, establishing the inevitability of entropy increase in isolated systems, captures the essence of time's unidirectional flow and its role in the universe's evolution. This section delves deeply into these concepts, elucidating their implications across a spectrum of physical and chemical phenomena.

Entropy, denoted by S, is a thermodynamic property representing the degree of disorder or randomness in a system. It is a state function, pivotal to determining a system's energy distribution at the microscopic level. The concept of entropy transcends simple disorder, providing a measure of how energy is dispersed or spread out in a process or transformation.

The formulation of entropy mathematically stems from Ludwig Boltzmann, who related entropy to the number of microscopic configurations (Ω) that correspond to a thermodynamic system's macroscopic state. This relationship is expressed as:

$$S = k \ln \Omega$$

where k is the Boltzmann constant (1.38×10^{-23} J/K) and Ω represents the number of possible configurations. The Boltzmann formula bridges microscopic statistical mechanics with macroscopic thermodynamics, providing a framework to analyze entropy in atomic and molecular contexts.

Consider a simple example of gas expansion within a volume—it proceeds from a confined region to fill the entire available space. As the gas molecules spread and occupy more configurations, the entropy increases, harnessing a pathway towards thermodynamic equilibrium.

This leads to the *second law of thermodynamics*, a fundamental construct stipulating that in any spontaneous process, the entropy of an isolated system will increase. Formally, for any irreversible process:

$$\Delta S_{\text{universe}} = \Delta S_{\text{system}} + \Delta S_{\text{surroundings}} > 0$$

For reversible processes, the entropy change can achieve zero, indicating full equilibrium. The second law highlights that the universe evolves naturally towards states of higher entropy, an insight central to understanding energy transformations and irreversibilities.

In practical terms, the second law introduces constructs such as *thermal efficiency* and the *Carnot cycle*, illustrating that no process converting heat to work, like a heat engine, can be completely efficient. The *Carnot theorem* defines the maximum efficiency achievable by a heat engine operating between two thermal reservoirs:

$$\eta = 1 - \frac{T_C}{T_H}$$

where T_C and T_H are the absolute temperatures of the cold and hot reservoirs, respectively. The Carnot cycle provides an idealized but theoretical limit for the efficiency of real-world heat engines, serving as a benchmark to improve thermal systems.

Entropy's significance extends to chemical reactions, where it serves as a deciding factor in reaction spontaneity. The *Gibbs free energy* (G), a pivotal state function, encapsulates this relationship at constant temperature and pressure:

$$\Delta G = \Delta H - T\Delta S$$

Here, ΔG determines reaction spontaneity ($\Delta G < 0$), linking enthalpy ΔH, temperature T, and entropy change ΔS in a harmonized framework. For reactions at equilibrium, $\Delta G = 0$, allowing the calculation of reaction dynamics when entropy and enthalpy changes are known.

For illustration, consider a calorimetric approach to melting ice:

$$H_2O \text{ (s)} \longrightarrow H_2O \text{ (l)}$$

Here, $\Delta H > 0$ (endothermic reaction), but the significant increase in entropy ($\Delta S > 0$) ensures $\Delta G < 0$ above $0\,°C$, classifying the process as spontaneous at those temperatures.

Entropy also provides the basis for understanding *degeneracy*, particularly within systems demonstrating quantum states. Systems allocate energy across available quantum states, leading to an understanding of configurations that maximize randomness, supported by the Boltzmann definition.

Significant insights arise in statistical mechanics from the *entropy of mixing*, essential when different gases intermingle. The entropy change associated

4.4. ENTROPY AND THE SECOND LAW

with such mixing processes is evaluated by:

$$\Delta S_{\mathrm{mix}} = -nR \sum_i x_i \ln x_i$$

where n is the total number of moles, R is the ideal gas constant, and x_i represents the mole fraction of each component. This calculation underscores the entropy increase inherent in mixing disparate substances, forming the basis for understanding solutions and chemical mixtures.

Moreover, the second law governs natural phenomena outside isolated systems, impacting biological processes, ecological systems, and even cosmological models. For instance, life sustains low entropy—even if temporarily—by intensively exchanging matter and energy with its environment, an open-system reformulation of second-law principles.

At an ecological level, thermodynamics aids in quantifying energy efficiency and entropy production within ecosystems, modeling the sustainability and health of environmental systems. These frameworks inform conservation strategies and technological adaptations to conserve resources and minimize ecological impacts.

The concept of *entropy production* elucidates dissipation processes, where any transport phenomenon or reaction is accompanied by entropy generation. This forms a cornerstone in irreversible thermodynamics, contributing to non-equilibrium states' theoretical modeling in contexts ranging from chemical kinetics to materials science.

Entropy and the second law also achieve metaphysical dimensions, emblematic of life and time, reflecting existential inquiries into the universe's fate—commonly captured within the *heat death* hypothesis outlining ultimate thermodynamic equilibrium devoid of usable energy.

Overall, entropy and the second law's insights manifest in myriad applications, from exploring energy efficiency in engineered systems to unraveling complex biological and ecological structures' energetics. They form an integral part of the thermodynamic narrative, tracing the irreversible journey from order to disorder, encapsulating the essence of natural progression beyond the immediacy of observable phenomena. The ramifications extend through scientific disciplines, embodying a unifying principle impacting technological, natural, and philosophical inquiry.

4.5 Free Energy and Chemical Potential

In thermodynamics, the concepts of free energy and chemical potential are instrumental in understanding the spontaneity and equilibrium of reactions and processes. These constructs offer a deeper insight into how systems navigate energy distributions, react under constraints, and transition between phases or states. Free energy quantifies the capacity for work within a system, while chemical potential provides the driving force for the distribution and movement of particles. Together, they form a robust framework that guides the analysis of chemical equilibria and phase transitions in diverse scientific disciplines.

Free energy encompasses several related thermodynamic potentials, with Gibbs free energy (G) and Helmholtz free energy (A) being the most prevalent. Each potential serves a different purpose, depending on the external constraints imposed on the system.

Gibbs free energy (G), used predominantly at constant temperature and pressure, is defined as:

$$G = H - TS$$

where H represents enthalpy, T is absolute temperature, and S is entropy. Gibbs free energy provides a criterion for reaction spontaneity and equilibrium. The change in Gibbs free energy (ΔG) allows us to predict whether a process will proceed spontaneously:

- $\Delta G < 0$: the process is spontaneous.
- $\Delta G > 0$: the process is non-spontaneous.
- $\Delta G = 0$: the system is at equilibrium.

Gibbs free energy is central to chemical reactions, phase changes, and biochemical pathways, reflecting the balance between enthalpic and entropic contributions to a system's evolution.

Consider a chemical reaction:

$$aA + bB \longrightarrow cC + dD$$

4.5. FREE ENERGY AND CHEMICAL POTENTIAL

The change in free energy for this reaction (ΔG) is linked to the reactants' and products' standard free energies of formation (G_f°) by the relationship:

$$\Delta G = \sum_{products} nG_f^\circ - \sum_{reactants} mG_f^\circ$$

The relationship between ΔG and the equilibrium constant K of a reaction is captured in the equation:

$$\Delta G = \Delta G^\circ + RT \ln Q$$

where ΔG° is the standard Gibbs free energy change, R is the universal gas constant, T is the absolute temperature, and Q is the reaction quotient. At equilibrium, $Q = K$ and $\Delta G = 0$, leading to the expression:

$$\Delta G^\circ = -RT \ln K$$

This equation fundamentally links the thermodynamic stability of a reaction to the concentrations of reactants and products.

Helmholtz free energy (A), defined as:

$$A = U - TS$$

where U denotes internal energy, is equally critical, particularly for processes occurring at constant volume and temperature. It quantifies the maximum reversible work obtainable from a system excluding expansion work:

$$dA = -SdT - PdV$$

Helmholtz energy becomes relevant in systems like confined gases where pressure-volume work is limited.

Transitioning to the concept of *chemical potential* (μ), it denotes the change in free energy with the addition of particles to a system at constant temperature and pressure. In simpler terms, it represents the potential energy stored within particles that can drive changes in composition:

$$\mu = \left(\frac{\partial G}{\partial n_i}\right)_{T,P,n_{j\neq i}}$$

where n_i is the number of moles of component i.

Chemical potential serves as a fundamental criterion for phase equilibria, influencing phenomena such as diffusion, electrochemical reactions, and phase transitions. For phase equilibrium between two phases α and β, the chemical potentials must equalize:

$$\mu_i^\alpha = \mu_i^\beta$$

This requirement dictates conditions where phases coexist, guiding the interpretation of phase diagrams.

Consider the example of a liquid-vapor equilibrium:

$$H_2O\ (l) \rightleftharpoons H_2O\ (g)$$

At the boiling point, $\mu_{H2O(l)} = \mu_{H2O(g)}$, driving the temperature dependence known as the Clausius-Clapeyron relation:

$$\frac{dP}{dT} = \frac{\Delta H_{vap}}{T\Delta V}$$

where ΔH_{vap} is the molar enthalpy of vaporization and ΔV is the change in molar volume.

Chemical potential also reveals insight into systems under electrochemical conditions. In an electrochemical cell, for instance, electric work done relates to the change in Gibbs free energy, infused with an electrical component:

$$\Delta G = -nFE$$

where n is the number of moles of electrons transferred, F is Faraday's constant, and E is the electromotive force of the cell. This relation forms the backbone of electrochemical thermodynamics, guiding applications from battery technology to corrosion prevention.

4.5. FREE ENERGY AND CHEMICAL POTENTIAL

The change in free energy for this reaction (ΔG) is linked to the reactants' and products' standard free energies of formation (G_f°) by the relationship:

$$\Delta G = \sum_{products} nG_f^\circ - \sum_{reactants} mG_f^\circ$$

The relationship between ΔG and the equilibrium constant K of a reaction is captured in the equation:

$$\Delta G = \Delta G^\circ + RT \ln Q$$

where ΔG° is the standard Gibbs free energy change, R is the universal gas constant, T is the absolute temperature, and Q is the reaction quotient. At equilibrium, $Q = K$ and $\Delta G = 0$, leading to the expression:

$$\Delta G^\circ = -RT \ln K$$

This equation fundamentally links the thermodynamic stability of a reaction to the concentrations of reactants and products.

Helmholtz free energy (A), defined as:

$$A = U - TS$$

where U denotes internal energy, is equally critical, particularly for processes occurring at constant volume and temperature. It quantifies the maximum reversible work obtainable from a system excluding expansion work:

$$dA = -SdT - PdV$$

Helmholtz energy becomes relevant in systems like confined gases where pressure-volume work is limited.

Transitioning to the concept of *chemical potential* (μ), it denotes the change in free energy with the addition of particles to a system at constant temperature and pressure. In simpler terms, it represents the potential energy stored within particles that can drive changes in composition:

$$\mu = \left(\frac{\partial G}{\partial n_i}\right)_{T,P,n_{j\neq i}}$$

where n_i is the number of moles of component i.

Chemical potential serves as a fundamental criterion for phase equilibria, influencing phenomena such as diffusion, electrochemical reactions, and phase transitions. For phase equilibrium between two phases α and β, the chemical potentials must equalize:

$$\mu_i^\alpha = \mu_i^\beta$$

This requirement dictates conditions where phases coexist, guiding the interpretation of phase diagrams.

Consider the example of a liquid-vapor equilibrium:

$$\text{H}_2\text{O (l)} \rightleftharpoons \text{H}_2\text{O (g)}$$

At the boiling point, $\mu_{H2O(l)} = \mu_{H2O(g)}$, driving the temperature dependence known as the Clausius-Clapeyron relation:

$$\frac{dP}{dT} = \frac{\Delta H_{vap}}{T\Delta V}$$

where ΔH_{vap} is the molar enthalpy of vaporization and ΔV is the change in molar volume.

Chemical potential also reveals insight into systems under electrochemical conditions. In an electrochemical cell, for instance, electric work done relates to the change in Gibbs free energy, infused with an electrical component:

$$\Delta G = -nFE$$

where n is the number of moles of electrons transferred, F is Faraday's constant, and E is the electromotive force of the cell. This relation forms the backbone of electrochemical thermodynamics, guiding applications from battery technology to corrosion prevention.

Furthermore, free energy assumes critical significance in *biochemical systems*, where it shapes metabolic pathways and enzyme actions. Enzymatic reactions, facilitated by the lowering of activation energies, depend on Gibbs free energy alterations. The coupling of endergonic ($\Delta G > 0$) and exergonic ($\Delta G < 0$) reactions underpins ATP's role as an energy currency within cellular environments, facilitating biological work processes:

$$ATP + H_2O \longrightarrow ADP + P_i$$

introducing a pivotal energy-producing reaction in cellular energetics.

Finally, the *Le Chatelier's principle* finds its basis within the realm of free energy, postulating that systems counteract applied perturbations to restore equilibrium. Quantifying these shifts and predicting new equilibrium states require a sophisticated understanding of how free energy terms balance each other to accommodate changes in pressure, temperature, or concentration.

Overall, free energy and chemical potential embody essential principles in thermodynamics, bridging empirical observations with theoretical models. They elucidate interactions from atomic-molecular scales to macroscopic systems, portraying a universe governed by energetic compulsions towards equilibrium and uniformity. These concepts feed into technological advancements and scientific exploration, resolving complexities in energy engineering, material science, environmental implications, and life-sustaining processes, weaving a cohesive narrative that ties individual systems to universal dynamics.

4.6 Chemical Equilibrium and Le Chatelier's Principle

Chemical equilibrium represents a fundamental concept in chemistry, embodying the state where reactants and products coexist in a dynamic balance in reversible reactions. This section explores the principles governing chemical equilibrium and introduces Le Chatelier's Principle, a predictive tool for understanding how systems respond to changes in external conditions. Together, these constructs provide a comprehensive framework to analyze reactions' behavior, optimize industrial processes, and predict shifts in system states under varying environmental stresses.

Chemical equilibrium is defined as the state in a reversible reaction where the rate of the forward reaction equals the rate of the backward reaction, leading to constant concentrations of reactants and products over time. At equilibrium, although macroscopic properties appear static, molecular-level reactions continue to occur dynamically—a phenomenon known as dynamic equilibrium.

For a general reversible reaction:

$$aA + bB \rightleftharpoons cC + dD$$

The equilibrium constant (K) quantifies the ratio of product concentrations to reactant concentrations at equilibrium:

$$K_c = \frac{[C]^c[D]^d}{[A]^a[B]^b}$$

where $[A]$, $[B]$, $[C]$, and $[D]$ are the molar concentrations of reactants and products. The equilibrium constant (K_c for concentrations, K_p for partial pressures) is dependent on temperature but remains unaffected by changes in concentrations or pressures of reactants and products.

Understanding K provides insight into the reaction mixture's composition at equilibrium:

- $K \gg 1$ suggests product formation is favored at equilibrium.
- $K \ll 1$ implies that the reactants predominate at equilibrium.
- $K \approx 1$ indicates significant concentrations of both reactants and products at equilibrium.

The equilibrium state is not isolated from external influences. *Le Chatelier's Principle* provides a predictive framework describing how a chemical system at equilibrium responds to external perturbations, such as changes in concentration, pressure, or temperature. According to Le Chatelier's Principle, if a dynamic equilibrium is disturbed by changing the conditions, the position of equilibrium shifts to counteract the imposed change.

- **Effect of Concentration Changes**: If the concentration of a reactant or product in an equilibrated system is changed, the equilibrium position shifts to reduce the effect of the change. For instance, adding more

4.6. CHEMICAL EQUILIBRIUM AND LE CHATELIER'S PRINCIPLE

reactant A to the above reaction will shift the equilibrium toward the right, favoring product formation.

For example, consider the reversible reaction:

$$N_2\,(g) + 3\,H_2\,(g) \rightleftharpoons 2\,NH_3\,(g)$$

Increasing the concentration of H_2 forces the equilibrium to shift rightward, favoring the production of NH_3.

- **Effect of Pressure Changes**: For reactions involving gases, changes in pressure influence equilibrium. According to Le Chatelier's Principle, increasing the pressure will shift the equilibrium toward the side with fewer moles of gas, thereby reducing pressure.

  Reconsidering the $N_2 + 3\,H_2 \rightleftharpoons 2\,NH_3$ reaction, increasing the pressure will favor the formation of NH_3, as the product side contains fewer moles of gas than the reactant side.

- **Effect of Temperature Changes**: Temperature changes shift equilibrium depending on the endothermic or exothermic nature of the reaction. For endothermic reactions ($\Delta H > 0$), increasing temperature shifts equilibrium to the right (product-favored). Conversely, exothermic reactions ($\Delta H < 0$) shift leftward with a temperature increase.

  For example, in the exothermic reaction:

  $$2\,SO_2\,(g) + O_2\,(g) \rightleftharpoons 2\,SO_3\,(g)$$

  Raising the temperature displaces the equilibrium toward the reactants, as the system absorbs added heat by favoring the endothermic reverse reaction.

- **Effect of Catalysts**: Catalysts alter the rate of reaction but not the equilibrium position. They equally lower the activation energy for both directions of the reaction, thereby achieving equilibrium more rapidly without affecting the equilibrium composition.

From a practical standpoint, Le Chatelier's Principle finds significant utility in industrial applications such as the Haber process for ammonia synthesis or the

Contact process for sulfuric acid production. By manipulating reaction conditions (temperature, pressure, and concentration), industries optimize yields and efficiency, enhancing economic viability.

For example, the Haber process for producing ammonia (NH_3) involves:

$$N_2\,(g) + 3\,H_2\,(g) \rightleftharpoons 2\,NH_3\,(g)$$

Operating under high pressures and moderate temperatures, with the aid of catalysts (iron-based), aligns with thermodynamic and kinetic considerations to favor maximum ammonia yield within economic limits.

Equilibria and Le Chatelier's Principle also extend their relevance to biochemical systems, where cellular conditions influence metabolic pathways and enzymatic reactions. Living systems harness these principles, maintaining homeostasis by shifting equilibria to counter instructional signals, thereby regulating the synthesis and breakdown of biomolecules.

The intricate balance between an organism's internal and external environments exemplifies a natural application of Le Chatelier's Principle, where feedback mechanisms adjust metabolic pathways to ensure stability and adaptability.

In ecological contexts, equilibrium concepts help predict and manage ecosystem dynamics, considering variables such as pollution levels, resource availability, and environmental changes. Remediation strategies in environmental sciences leverage equilibrium shifts to restrict pollutant levels, optimize remediation processes, and restore ecological balance.

Finally, Le Chatelier's Principle extends into educational paradigms, providing a foundational framework in academic curricula across chemistry disciplines. Practically, it aids in solving equilibrium problems, predicting shifts, and optimizing conditions for maximum productivity.

Quantitative treatments of chemical equilibrium involve the application of the *reaction quotient* Q, analogous to the equilibrium constant expression, evaluated at any point in the reaction:

$$Q = \frac{[C]^c[D]^d}{[A]^a[B]^b}$$

Comparing Q with K provides insight into reaction directionality:

- $Q < K$: The reaction will progress forward until equilibrium is reached.

- $Q > K$: The reaction shifts backward to attain equilibrium.

- $Q = K$: The system is at equilibrium, with no net change observed.

This quantitative analysis forms the backbone of computational tools used to predict equilibrium positions in complex scenarios, facilitating advanced research in material science, chemical engineering, and pharmaceutical development.

Collectively, chemical equilibrium and Le Chatelier's Principle offer profound insights into reaction dynamics, guiding innovations in chemical synthesis, process optimization, and environmental management. These principles not only elucidate foundational chemistry but also inspire interdisciplinary solutions to contemporary scientific challenges.

4.7 Phase Equilibria and Phase Diagrams

Phase equilibria and phase diagrams are critical elements of thermodynamics that describe the conditions under which distinct phases of matter coexist in equilibrium. These constructs provide a comprehensive framework for understanding phase transitions, the behavior of substances under varying conditions of temperature and pressure, and the resulting phase distributions. By exploring phase equilibria and their graphical representations through phase diagrams, we gain insight into material properties, predict responses to environmental changes, and design processes in chemistry, metallurgy, and materials science.

The concept of *phase* pertains to a distinct homogeneous region within a system that differs in structure, composition, or state from other regions of the system. Common states of matter include solid, liquid, and gas, but numerous other phases may exist depending on molecular arrangement and interactions.

Phase equilibrium occurs when multiple phases coexist at thermodynamic equilibrium, with no net transfer of material between them. This equilibrium results from a dynamic balance where phase changes such as melting, vaporization, or sublimation occur at equal rates, maintaining stable macroscopic properties.

Understanding the conditions that foster phase equilibria involves two central principles:

1. **Gibbs Phase Rule**: The Gibbs Phase Rule provides a formula to predict the degrees of freedom (variance) in a system at equilibrium in terms of the number of phases (P) and components (C):

$$F = C - P + 2$$

Here, F signifies the number of independent variables (such as temperature or pressure) that can be altered without changing the number of phases present. A reduced F implies extensive constraints on experimental or natural conditions.

As an example, consider a single-component system like water ($C = 1$), where equilibrium between three phases (solid, liquid, gas) is characterized by the *triple point*, a condition where $F = 0$, indicating that temperature and pressure are uniquely fixed.

2. **Clapeyron Equation**: The Clapeyron equation offers a thermodynamic description of phase boundaries, relating the enthalpy of phase transition (ΔH) to changes in temperature and pressure:

$$\frac{dP}{dT} = \frac{\Delta H}{T \Delta V}$$

This equation is pivotal in understanding how phase boundaries shift with changing conditions, underpinning the design of phase diagrams and guiding predictions of phase stability.

Phase diagrams serve as visual representations of the equilibrium between phases under varying conditions. They provide a map indicating regions of phase stability, phase transition lines, critical points, and triple points, encapsulating complex thermodynamic data succinctly.

- **Single-Component Systems**: The phase diagram of a single-component system, such as water, highlights critical temperature-pressure relationships. A typical water phase diagram showcases:
    - *Fusion Line*: Separates solid and liquid phases, indicating melting or freezing points.
    - *Vaporization Line*: Demarcates liquid and vapor regions, defining boiling points.

4.7. PHASE EQUILIBRIA AND PHASE DIAGRAMS

- *Sublimation Line*: Distinguishes solid and vapor phases, pertinent in low-pressure environments.
- *Triple Point*: The unique temperature and pressure where solid, liquid, and vapor coexist in balance.
- *Critical Point*: The end-point of the vaporization line where the distinction between liquid and gas ceases.

- **Binary and Multi-Component Systems**: In binary systems, phase diagrams evolve in complexity, as they must consider compositional variations along with temperature and pressure changes. Such diagrams typically feature:
  - *Isotherms and Isobars*: Lines of constant temperature or pressure, indicating shifting phase equilibria with composition changes.
  - *Liquidus and Solidus Lines*: Boundaries depicting regions of complete liquid or solid phases, crucial in alloy systems, guiding the development of metallic structures with varied compositional blends.
  - *Eutectic and Peritectic Points*: Characteristic features where phases converge under specific conditions, dictating alloy solidification processes or compound formation.

Applications of phase diagrams span numerous disciplines:

- **Metallurgy and Material Science**: Phase diagrams are indispensable tools in metallurgical engineering and materials science. They enable precise control over the alloying processes, optimizing mechanical properties such as strength, ductility, and corrosion resistance. For instance, in the iron-carbon system, phase diagrams predict phases like austenite or cementite, informing methods to enhance steel's workability or hardness through heat treatment processes.

- **Chemical Engineering and Process Design**: In chemical engineering, phase diagrams facilitate reactor design and separation processes, like distillation or crystallization, by elucidating phase equilibria. This knowledge enables optimized reaction conditions, maximizing yield and efficiency, particularly in industries reliant on complex reaction networks or multi-phase reactions.

- **Earth and Planetary Sciences**: Phase diagrams extend into geochemistry and planetology, modeling the behavior of minerals under extreme pressures and temperatures encountered in Earth's mantle or other planetary interiors. This allows for a deeper understanding of geological events, enabling interpretations of seismic data or extrapolations concerning the composition of extraterrestrial bodies.

- **Pharmaceuticals and Chemistry**: In the pharmaceutical domain, phase diagrams predict solubility and stability of compounds, guiding formulation strategies for drugs exhibiting polymorphism. This ensures therapeutic efficacy, safety, and control over drug release mechanisms.

Incorporating *solution thermodynamics*, phase diagrams elaborate phase behaviors in liquid mixtures through activity coefficients or fugacities, extending concepts from ideal to real solutions. This encapsulates deviations from Raoult's or Henry's laws, critical for modeling non-ideal interactions in concentrated or dilute systems—pertinent in extraction processes, equilibrium modeling, or gas solubility studies.

Furthermore, experiments such as *Differential Scanning Calorimetry (DSC)* and *Thermogravimetric Analysis (TGA)* quantify phase transitions' thermal properties, validating phase diagram predictions practically. These techniques measure endothermic or exothermic transitions, tracking mass changes across broad conditions to depict materials' thermal responses.

Finally, *critical phenomena* related to phase equilibria introduce insights into phase stability and responses to perturbations beyond conventional conditions. Exploring the *supercritical fluid state*, characterized by properties of both liquids and gases, facilitates advances in solvent applications, offering unique advantages in extraction or chemical synthesis without traditional solvents' constraints.

Through enriched understanding, phase equilibria and diagrams underpin advancements across scientific and engineering disciplines, driving innovations and efficiencies that extend to metal alloys, chemical manufacturing, environmental technologies, and pharmaceutical developments. They not only serve as educational cornerstones in thermodynamics but also embody a quintessential link connecting theoretical predictions and empirical realities, translating complex molecular interactions into accessible, actionable insights.

Chapter 5

Kinetics and Reaction Dynamics

The study of kinetics and reaction dynamics unveils the temporal progression and mechanistic intricacies of chemical reactions, providing insight into the forces and factors that influence reaction rates. This chapter delves into the quantitative characterization of reaction kinetics, exploring rate laws and the impact of temperature, pressure, and concentration. Through detailed examination of reaction mechanisms, including the role of transition states and intermediates, the chapter elucidates how complex reactions are navigated and understood. Catalysis, both chemical and enzymatic, is addressed for its transformative role in modulating reaction pathways, offering insights into the strategic acceleration of chemical processes. Together, these topics compose a cohesive framework for understanding the kinetic control and dynamic behavior of chemical systems.

5.1 Basic Principles of Chemical Kinetics

Chemical kinetics is the branch of chemistry that studies the speed or rate at which chemical reactions occur. Understanding reaction rates and how they can be controlled is fundamental to the design of new reactions, the synthesis of products, and the optimization of processes in chemical manufacturing, bi-

CHAPTER 5. KINETICS AND REACTION DYNAMICS

ology, and environmental science. The basic principles of chemical kinetics serve as the foundation for more complex concepts in reaction dynamics and thermodynamics.

The study of reaction rates involves several key components: the concentration of reactants, the temperature of the system, and the presence of catalysts or inhibitors. These factors are interrelated, influencing the speed at which molecules collide and initiate chemical change.

Reaction Rate and its Measurement

The rate of a chemical reaction is defined as the change in concentration of reactants or products per unit time. It is usually expressed in terms of molarity per second (mol/L/s). This rate can be calculated by measuring how the concentration of a reactant decreases over time or how the concentration of a product increases.

Mathematically, the average rate of a reaction $A \to B$ can be expressed as:

$$\text{Rate}_{avg} = -\frac{\Delta[A]}{\Delta t} = \frac{\Delta[B]}{\Delta t}$$

Where $[A]$ and $[B]$ are the concentrations of reactant A and product B, respectively. The negative sign for the reactant indicates that its concentration decreases over time.

The instantaneous rate of reaction, which is more physically meaningful, can be determined by taking the derivative of the concentration with respect to time:

$$\text{Rate}_{inst} = -\frac{d[A]}{dt} = \frac{d[B]}{dt}$$

When considering complex reactions involving multiple reactants and products, the reaction rate is often expressed in terms of a rate law.

Rate Laws and Reaction Orders

Rate laws express the relationship between the rate of a chemical reaction and the concentration of its reactants. A rate law has the general form:

$$\text{Rate} = k[A]^m[B]^n \ldots$$

5.1. BASIC PRINCIPLES OF CHEMICAL KINETICS

Here, k is the rate constant, and m and n are the reaction orders with respect to reactants A and B, respectively. These exponents are typically determined experimentally and can differ from the stoichiometric coefficients in the balanced chemical equation.

The overall order of a reaction is the sum of the orders with respect to each reactant. It provides insight into the reaction mechanism, suggesting the number of molecular interactions in the rate-determining step.

Factors Influencing Reaction Rates

Several key factors control the speed of a chemical reaction:

- **Concentration:** The rate of reaction typically increases as the concentration of reactants increases. This is due to the increased number of molecular collisions, which are necessary to initiate the reaction.

- **Temperature:** Reaction rates generally increase with temperature. According to the Arrhenius equation, the rate constant k is related to the temperature T and the activation energy E_a:

$$k = Ae^{-E_a/RT}$$

  Here, A is the pre-exponential factor, R is the universal gas constant, and T is the temperature in Kelvin. Higher temperatures provide molecules with greater kinetic energy, enhancing the probability of overcoming the activation energy barrier.

- **Catalysts:** Catalysts are substances that increase reaction rates without being consumed in the reaction. They function by providing an alternative reaction pathway with a lower activation energy. Enzymes are a specific type of biological catalyst that are highly specific and efficient.

- **Surface Area:** For reactions involving solids, an increase in surface area (e.g., by powdering a solid) exposes more reactive sites and accelerates the reaction.

- **Pressure:** In reactions involving gases, increasing the pressure effectively increases the concentration of reactants, thus boosting the reaction rate.

An illustrative example is the reaction of magnesium metal with hydrochloric acid:

$$\text{Mg (s)} + 2\,\text{HCl (aq)} \longrightarrow \text{MgCl}_2\,\text{(aq)} + \text{H}_2\,\text{(g)}$$

In this reaction, increasing the concentration of hydrochloric acid or raising the temperature would result in a faster evolution of hydrogen gas.

Reaction Mechanism

Chemical reactions typically proceed through a series of elementary steps or stages. The mechanism of a reaction includes all the individual steps that lead from reactants to products. Each step represents a single molecular event and has its own rate law.

The overall reaction rate is determined by the slowest step in the sequence, known as the rate-determining step. Understanding the reaction mechanism is crucial for modifying and optimizing chemical processes.

Consider the decomposition of hydrogen peroxide, a reaction often catalyzed by iodide ions:

$$2\,\text{H}_2\text{O}_2\,\text{(aq)} \longrightarrow 2\,\text{H}_2\text{O (l)} + \text{O}_2\text{(g)}$$

This reaction involves multiple steps, one of which is the slow, rate-determining decomposition of an intermediate species.

Application of Kinetics in Industry

Knowledge of chemical kinetics is vital in industries that rely on chemical processes. It allows chemists and engineers to design processes that are efficient, safe, and cost-effective.

For instance, the Haber process for synthesizing ammonia:

$$\text{N}_2\,\text{(g)} + 3\,\text{H}_2\,\text{(g)} \longrightarrow 2\,\text{NH}_3\,\text{(g)}$$

is conducted under high pressure and temperature with an iron catalyst to maximize the production rate and yield of ammonia. Understanding kinetics enables the adjustment of these conditions to optimize the process.

In pharmaceuticals, reaction kinetics are critical to optimizing drug synthesis and ensuring the stability of active ingredients. The kinetics of drug degradation informs the shelf life and storage conditions of pharmaceuticals.

Theoretical Models of Reaction Rates

5.1. BASIC PRINCIPLES OF CHEMICAL KINETICS

To complement empirical methods, theoretical models have been developed to predict and explain reaction kinetics. One widely used model is the transition state theory, which assumes that a high-energy transition state exists as reactants are converted into products.

Another fundamental model is collision theory, which holds that molecules must collide with sufficient energy and proper orientation for a reaction to occur. While simplified, this model provides insight into the factors that influence reaction rates.

Both these theories contribute essential frameworks for understanding how and why reactions occur at the molecular level.

Experimental Determination of Reaction Rates

Determining reaction rates experimentally involves tracking changes in reactant or product concentrations over time. Common techniques include:

- **Spectroscopy:** Monitoring absorbance or emission spectra to determine concentrations.

- **Titration:** Periodically withdrawing samples and titrating to measure reactant or product concentrations.

- **Gas Evolution:** Measuring the volume or pressure change when a gaseous product forms.

These methods enable the construction of concentration vs. time plots, from which rate laws and constants can be derived. For accuracy, experiments are carefully controlled to isolate the effects of individual variables like temperature and concentration.

The basic principles of chemical kinetics provide a framework for understanding the temporal evolution of reactions. By measuring and modeling reaction rates, chemists gain insights into the nature of chemical processes, guiding the synthesis and optimization of new compounds and technologies. Through a combination of experimental data and theoretical models, the field of kinetics continues to evolve, offering deeper insights into the dynamic behavior of chemical systems.

5.2 Rate Laws and Reaction Order

Chemical kinetics revolves heavily around the understanding of rate laws and reaction orders, which form the mathematical framework that describes the dynamics of chemical reactions. These concepts are pivotal for identifying how various factors influence the speed of a reaction and for drawing inferences about the underlying reaction mechanisms. A rate law not only quantifies the rate of reaction in terms of the concentration of reactants but also provides insight into the stoichiometry and potential pathways the reaction might follow.

A rate law is expressed in the general form:

$$\text{Rate} = k[A]^m[B]^n \ldots$$

where k is the rate constant, and m and n represent the orders of the reaction with respect to reactants A and B, respectively. The overall order of the reaction is the sum of the individual orders, $m + n + \ldots$.

The rate law of a reaction can be determined experimentally and may include one or more reactants. The two primary types of rate laws are differential and integrated rate laws.

Differential Rate Law: Also known as the rate equation, this form of rate law shows how the rate depends on the concentration of reactants. It is used to derive integrated rate laws that reveal the change in concentrations over time.

Integrated Rate Law: This form is derived from the differential rate law and provides a direct relationship between the concentration of reactants and time. It helps in understanding how long a reaction takes to reach completion under specified conditions.

Reaction order is a critical parameter in chemical kinetics as it indicates the power to which the concentration of a reactant is raised in the rate law. Based on experimental observations, reaction orders can be classified as zero-order, first-order, second-order, or even fractional or mixed order in certain cases. It is important to note that the reaction order is not necessarily related to the stoichiometry of the reaction equation.

Zero-Order Reactions:

In zero-order reactions, the rate is independent of the concentration of reactants. This implies that the concentration of reactants does not influence the

5.2. RATE LAWS AND REACTION ORDER

rate of reaction. The differential rate law for a zero-order reaction is:

$$\text{Rate} = k$$

The integrated rate law for a zero-order reaction is given by:

$$[A]_t = [A]_0 - kt$$

where $[A]_0$ is the initial concentration and $[A]_t$ is the concentration of A at time t. Such reactions are typical in processes where a catalyst is saturated by reactants, limiting the speed to a constant rate.

An example of a zero-order reaction is the decomposition of ammonia on a platinum surface:

$$2\,NH_3\,(g) \longrightarrow N_2\,(g) + 3\,H_2\,(g)$$

This reaction proceeds at a constant rate irrespective of the concentration of ammonia.

First-Order Reactions:

For first-order reactions, the rate depends linearly on one reactant concentration. The differential rate law is expressed as:

$$\text{Rate} = k[A]$$

The integrated rate law takes the form:

$$\ln[A]_t = \ln[A]_0 - kt$$

or equivalently:

$$[A]_t = [A]_0 e^{-kt}$$

The concentration of the reactant decays exponentially over time. The half-life ($t_{1/2}$) for a first-order reaction, which is the time required for the concentration to decrease to half its initial value, is constant:

$$t_{1/2} = \frac{\ln 2}{k}$$

This constancy makes first-order kinetics particularly useful in radioactive decay and pharmacokinetics, where substances show exponential decay with time.

A pertinent example is the radioactive decay of certain isotopes such as ^{14}C, used in radiocarbon dating:

$$^{14}C \longrightarrow {}^{14}N + \beta^-$$

Second-Order Reactions:

Second-order reactions involve either two molecules of the same reactant or two different reactants. The differential rate law for a simple second-order reaction is:

$$\text{Rate} = k[A]^2$$

The integrated rate law is:

$$\frac{1}{[A]_t} = \frac{1}{[A]_0} + kt$$

For a second-order reaction involving two different reactants, the rate law is:

$$\text{Rate} = k[A][B]$$

In this situation, the integrated form is more complex and typically requires consideration of initial concentrations, often leading to complex kinetic studies or approximations.

The half-life for a second-order reaction, unlike first-order reactions, depends on the initial concentration:

$$t_{1/2} = \frac{1}{k[A]_0}$$

5.2. RATE LAWS AND REACTION ORDER

A classic example of a second-order reaction is the dimerization of nitrogen dioxide:

$$2\,NO_2\,(g) \longrightarrow N_2O_4\,(g)$$

Reactions that do not fit neatly into simple integer orders are termed mixed-order or complex reactions. These can show fractional orders, indicative of complex reaction pathways or mechanisms involving multiple steps or intermediates.

For example, a reaction with an experimental rate law:

$$\text{Rate} = k[A]^{0.5}[B]^2$$

exhibits a half-order with respect to A and a second-order with respect to B. Such kinetics are often observed in reactions involving adsorption processes on catalytic surfaces or in reactions with a chain mechanism involving radicals.

The determination of reaction order is a critical aspect of chemical kinetics studies and is usually conducted experimentally. Several methods are employed to establish the order of a reaction:

- **Method of Initial Rates:** This involves measuring the initial rate of reaction for different initial concentrations of reactants. By plotting initial rates against initial concentrations, the order with respect to each reactant can be determined from the slope.

- **Integrated Rate Law Method:** By fitting concentration vs. time data to various integrated rate equations, one can linearize the data in a graph, determining the reaction order based on which graph yields a straight line.

- **Half-life Method:** Particularly useful for first-order reactions, where the independence of half-life from initial concentrations can be exploited.

- **Isolation Method:** The concentration of all reactants except one is held constant, this allows for the determination of order with respect to a single reactant.

These experimental approaches are often complemented by sophisticated instruments such as spectrophotometers and chromatographs that monitor the reaction progress over time, facilitating accurate kinetic analyses.

In summary, the concepts of rate laws and reaction orders provide a robust framework for the exploration and understanding of chemical reactions. By articulating how reaction rates depend on concentrations, chemists can engineer processes and devise mechanistic models that accurately predict chemical behavior under various conditions. The insights gained bridge theoretical predictions with practical applications, driving innovations in diverse fields such as pharmaceuticals, materials science, and environmental chemistry.

5.3 Temperature and Reaction Rate: The Arrhenius Equation

Temperature is a fundamental factor influencing the kinetics of chemical reactions. Typically, an increase in temperature leads to an increase in the reaction rate. This relationship is quantitatively described by the Arrhenius equation, a powerful tool in chemical kinetics that relates the rate constant of a reaction to the absolute temperature and activation energy. Understanding this relationship is crucial for controlling reaction rates in industrial processes, biological systems, and various scientific applications.

The Arrhenius equation is expressed as:

$$k = Ae^{-E_a/(RT)}$$

where k is the rate constant, A is the pre-exponential factor or frequency factor, E_a is the activation energy, R is the universal gas constant, and T is the absolute temperature in Kelvin. This equation illustrates the exponential nature of the temperature dependence of the rate constant.

Activation energy (E_a) is the minimum energy that reacting species must possess for a reaction to occur. At the molecular level, it represents the energy barrier that must be overcome for reactants to be converted into products. This energy barrier is not a static threshold; rather, it is associated with the transition state, a high-energy, unstable configuration of atoms that represents an intermediate stage between reactants and products. This is depicted in a reaction coordinate diagram, which plots the potential energy of a system as a

5.3. TEMPERATURE AND REACTION RATE: THE ARRHENIUS EQUATION

function of its geometry or configuration along the reaction path.

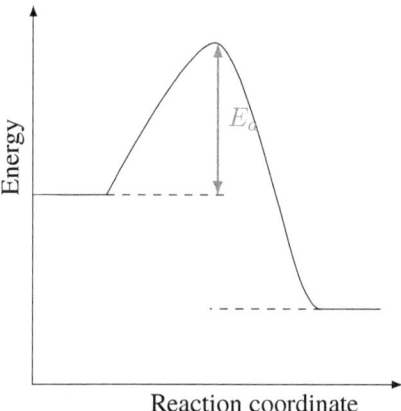

The diagram exemplifies that the activation energy is the energy difference between the energy of the transition state and the energy of the reactants. Higher activation energies imply slower reactions, as fewer molecules possess the requisite energy to surpass the barrier at a given temperature.

The pre-exponential factor A, also known as the frequency factor, is indicative of the frequency of collisions with the correct orientation for a reaction to occur. It encompasses several variables, including the number of effective collisions and the probability of correctly oriented collisions. While the Arrhenius equation highlights the influence of temperature and E_a, the pre-exponential factor accounts for steric factors and the intrinsic properties of reactants, such as molecular orientation and vibration.

The Arrhenius equation can be rearranged to a linear form suitable for graphical analysis:

$$\ln k = \ln A - \frac{E_a}{R} \cdot \frac{1}{T}$$

By plotting $\ln k$ versus $1/T$, a straight line is obtained, the slope of which is $-E_a/R$. The intercept of this line corresponds to $\ln A$. This method provides a straightforward means of determining the activation energy from experimental data.

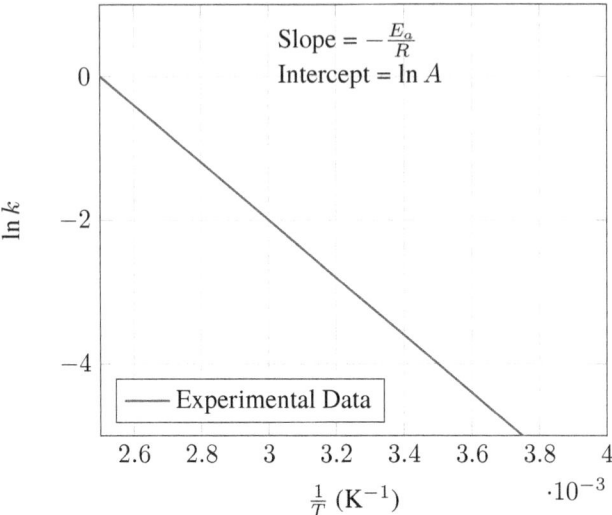

This visual representation elegantly conveys the inverse relationship between activation energy and reaction rate, showcasing that higher activation energies manifest as steeper negative slopes.

The temperature dependence of reaction rates underscores a key principle in chemistry: an increase in temperature typically results in a dramatic increase in the reaction rate. This phenomenon can be explained through kinetic molecular theory, which asserts that temperature is a measure of the average kinetic energy of molecules. As temperature rises, molecules move faster, increasing the frequency and energy of collisions. Consequently, more molecules surpass the activation energy barrier, leading to a higher rate of reaction.

The exponential factor $e^{-E_a/(RT)}$ in the Arrhenius equation precisely captures this effect, emphasizing that even a modest increase in temperature results in a significant increase in the proportion of molecules with sufficient energy to react.

Understanding the temperature dependence of reaction rates is crucial across various scientific and industrial domains:

- **Chemical Manufacturing:** In industries producing chemicals, polymers, and pharmaceuticals, controlling the temperature during reactions is vital for optimizing yields and minimizing side reactions. By manipu-

5.3. TEMPERATURE AND REACTION RATE: THE ARRHENIUS EQUATION

lating temperature, manufacturers can adjust reaction rates to maximize efficiency and selectivity.

- **Food Science:** The Arrhenius equation is used to model the spoilage rates of perishable foods, helping to establish storage conditions and shelf lives. Elevated temperatures accelerate spoilage reactions by increasing the reactivity of natural enzymes and microorganisms, whereas controlling temperature can prolong freshness.

- **Biological Systems:** In biology, enzyme-catalyzed reactions are sensitive to temperature changes, and deviations from optimal temperatures can impair metabolic pathways and homeostasis. The mild temperature dependence of enzymatic reactions, relative to non-catalyzed reactions, is attributable to the lowering of activation energy by the enzyme.

- **Atmospheric Chemistry:** Environmental reactions, such as the decomposition of pollutants, are considerably influenced by atmospheric temperature fluctuations. The Arrhenius equation aids in modeling these temperature dependencies, thereby informing climate models and environmental policies.

While the Arrhenius equation provides substantial insight into the temperature dependence of reaction rates, further refinements exist to address its limitations. For instance, the transition state theory extends the Arrhenius concept by incorporating molecular structure into the activation energy framework.

Additionally, temperature-dependent changes in A and deviations from linear Arrhenius behavior can arise in complex systems. Cases where the Arrhenius plot is nonlinear suggest temperature-dependent changes in the reaction mechanism or the involvement of multiple pathways with different rates.

The extended forms of the Arrhenius equation account for these nuances, allowing the investigation of reactions under diverse conditions:

$$k(T) = A(T)e^{-E_a(T)/(RT)}$$

Determining the activation energy experimentally involves measuring the rate constant k at different temperatures and using the Arrhenius plot as described earlier. This approach is practical across laboratory settings, enabling chemists to elucidate reaction mechanisms by revealing the energy barriers involved.

Experimental techniques such as calorimetry, spectroscopy, and chromatography are employed to gather kinetic data over a range of temperatures. These data drive insights into reaction specificity and pathways, supporting the development of kinetic models that accurately describe observed behaviors.

To illustrate the application of Arrhenius concepts, consider the decomposition of hydrogen peroxide (H_2O_2):

$$2\,H_2O_2\,(aq) \longrightarrow 2\,H_2O\,(l) + O_2\,(g)$$

This reaction is a classical example with Arrhenius-type temperature dependence. The decomposition accelerates with an increase in temperature as the barriers posed by the activation energy are more easily overcome. A catalyzed version, using potassium iodide as a catalyst, further exemplifies how E_a can be lowered by catalytic intervention, altering kinetics to achieve desired rates at milder temperatures.

The Arrhenius plot for this decomposition highlights the linear regime expected for such simple systems, providing both the activation energy and pre-exponential factor values relevant to understanding and optimizing reaction conditions.

The Arrhenius equation remains a cornerstone of chemical kinetics, revealing critical insights into the influence of temperature on reaction rates. Its applications extend into diverse fields, underscoring the universal value of understanding energy barriers and molecular collisions in dictating the pace of chemical transformations. The integration of Arrhenius principles with advanced modeling and experimental techniques continues to drive innovations and deepen our comprehension of chemical processes in both controlled and natural environments.

5.4 Reaction Mechanisms and the Steady-State Approximation

In chemical kinetics, understanding the mechanism behind a reaction is as crucial as understanding its rate. A reaction mechanism provides a detailed description of the individual steps that occur as reactants are transformed into products. Each step, called an elementary reaction, involves a simple interaction between molecules, and collectively, these steps comprise the overall

5.4. REACTION MECHANISMS AND THE STEADY-STATE APPROXIMATION

reaction. Real-world reactions often involve multi-step mechanisms, where the steps occur sequentially or simultaneously, leading to the formation of intermediates.

The elucidation of reaction mechanisms is an essential aspect of chemistry because it uncovers explanatory insights into the pathways and transformations that take place at a molecular level. This knowledge allows chemists to manipulate reactions to improve efficiency, control product distribution, and reduce unwanted side reactions.

An elementary reaction is a single-step process where reactants directly convert into products. The molecularity of an elementary reaction indicates the number of molecules coming together to react in that step and is categorized as unimolecular, bimolecular, or termolecular, depending on whether one, two, or three entities are involved.

- **Unimolecular Reactions:** These involve the transformation of a single reactant molecule into products. For example, the isomerization of cyclopropane (C_3H_6) to propene (C_3H_6) is a unimolecular reaction.

$$C_3H_6\,(g) \longrightarrow C_3H_6\,(g)$$

- **Bimolecular Reactions:** These occur when two reactant molecules collide to form products. A common example is the reaction between nitrogen dioxide (NO_2) and carbon monoxide (CO) to form nitrogen monoxide (NO) and carbon dioxide (CO_2).

$$NO_2\,(g) + CO\,(g) \longrightarrow NO\,(g) + CO_2\,(g)$$

- **Termolecular Reactions:** These involve three molecules colliding simultaneously, which is a rare event due to the low probability of three molecules colliding with the correct orientation and energy.

The overall reaction order for an elementary step is equal to its molecularity, as the rate law can be directly written from the stoichiometry of the elementary reaction.

For a multi-step reaction, the overall rate law is determined by its mechanism. A simple rate law cannot always be deduced from the stoichiometric equation because it may overlook intermediates formed or consumed during the reaction. Instead, it is often dominated by the slowest step in the mechanism, known as the rate-determining step (RDS).

The RDS imposes a kinetic bottleneck on the entire process, controlling the reaction rate. By analyzing the RDS, chemists can obtain information about reaction kinetics and conditions that maximize efficiency.

For example, consider the following two-step reaction mechanism for the formation of NO_2 from NO and O_2:

- $2\,NO\,(g) \longrightarrow N_2O_2\,(g)$ (fast equilibrium)
- $N_2O_2\,(g) + O_2\,(g) \longrightarrow 2\,NO_2\,(g)$ (slow step)

Here, N_2O_2 is an intermediate. The slow second step determines the rate:

$$\text{Rate} = k[N_2O_2][O_2]$$

The concentration of the intermediate $[N_2O_2]$ is not measurable directly and must be expressed in terms of the reactants using the equilibrium established in the fast step:

$$K[N_2O_2] = [NO]^2$$

where K is the equilibrium constant for the fast step.

In complex mechanisms, where intermediates are present, direct computation of reaction rates can become challenging. The steady-state approximation provides a simplifying assumption that facilitates this calculation. This approximation assumes that the concentration of an intermediate remains relatively constant through most of the reaction, reached rapidly after the initiation and before the system reaches equilibrium.

Mathematically, under steady-state conditions, the rate of change of the concentration of the intermediate is:

$$\frac{d[\text{Intermediate}]}{dt} \approx 0$$

This condition allows for the simplification of reaction rate equations, ensuring only slowly changing concentrations influence the rate of the overall reaction.

5.4. REACTION MECHANISMS AND THE STEADY-STATE APPROXIMATION

For example, in the reaction mechanism:

- $A + B \longrightarrow C$ (fast)
- $C \longrightarrow D$ (slow)

For intermediate C, the steady-state expression becomes:

$$\frac{d[C]}{dt} = k_1[A][B] - k_2[C] \approx 0$$

Solving this expression for $[C]$ yields:

$$[C] = \frac{k_1[A][B]}{k_2}$$

Replacing $[C]$ in the rate law for the slow step, the overall rate law becomes:

$$\text{Rate} = k_2[C] = \frac{k_1 k_2 [A][B]}{k_2} = k_1[A][B]$$

Thus, the steady-state approximation allows the construction of a rate law that reflects the contribution of the earliest steps, providing insights into the mechanism.

Understanding reaction mechanisms and employing the steady-state approximation are key aspects of various scientific endeavors:

- **Catalysis:** Catalytic processes frequently proceed through complex mechanisms involving multiple steps and transient intermediates. In industrial catalysis, such as the hydrogenation of ethene (C_2H_4) to ethane (C_2H_6) using nickel, the reaction involves the adsorption of H_2 and C_2H_4 on the catalyst surface. By modeling these steps through the steady-state approximation, it's possible to optimize catalytic conditions and improve reaction yields.

- **Atmospheric Chemistry:** Many reactions within the Earth's atmosphere—such as those responsible for ozone (O_3) formation and depletion—involve radical intermediates with very short lifetimes. Mechanistic modeling allows for quantifying how pollutants affect atmospheric reactivity, helping to inform regulatory policies and ozone management strategies.

- **Enzyme Kinetics:** Biological reactions catalyzed by enzymes often follow complex pathways with several intermediate stages. In the Michaelis-Menten mechanism for enzyme-substrate interactions, the steady-state approximation is employed to derive the Michaelis-Menten equation. This model describes how the reaction rate depends on substrate concentration:

$$v_0 = \frac{V_{\max}[S]}{K_m + [S]}$$

where v_0 is the initial reaction rate, $V_{\max}$ is the maximum rate, $[S]$ is the substrate concentration, and K_m is the Michaelis constant, reflective of enzyme affinity and catalytic efficiency.

With advancements in computational chemistry, molecular modeling and simulation have become integral to mechanism elucidation. Techniques such as molecular dynamics and quantum mechanics/molecular mechanics (QM/MM) simulations establish reaction pathways, exploring potential energy surfaces and providing detailed visualization of elementary processes. These approaches are synergistic with experimental methods, furnishing a comprehensive understanding of reactions.

Diagrammatic representation:

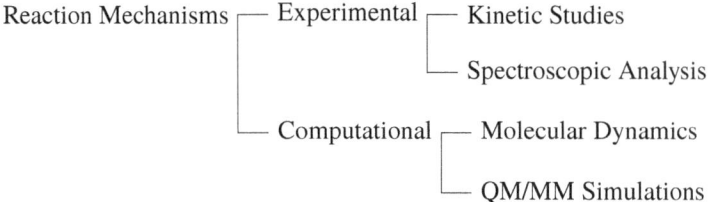

Identifying and validating mechanistic pathways involve integrating various experimental techniques:

- **Kinetic Isotope Effect (KIE):** This involves substituting atoms in a reactant with their isotopes—such as hydrogen with deuterium—and observing changes in reaction rates. Large KIE values suggest that the chemical bonds to the isotopically labeled atoms participate in the rate-determining step, providing clues to the mechanism.

- **Intermediate Detection Techniques:** Use of techniques such as NMR, mass spectrometry, and infrared spectroscopy to directly detect and characterize transient intermediates involved in the reaction.

- **Reaction Inhibition Studies:** Employing inhibitors to strategically block specific pathways and observe resulting changes in rate laws, thus providing insights to differentiate between parallel and sequential mechanisms.

The complex interplay between reactants through multi-step reaction mechanisms demands a detailed understanding to accurately predict and control chemical processes. The steady-state approximation stands as a pivotal theory simplifying the mathematical treatment of intermediates, aiding in the derivation of rate laws reflective of experimental realities. Through both theoretical modeling and empirical studies, the intricate pathways of reaction mechanisms can be dissected and optimized, enhancing numerous scientific, industrial, and biological operations. As science progresses, the confluence of experimental and computational approaches will continue to unravel the mechanistic intricacies governing chemical transformation.

5.5 Catalysis and Enzyme Kinetics

Catalysis is a fundamental concept in chemistry, playing an indispensable role in accelerating chemical reactions by lowering the activation energy required for the process. Catalysts are substances that, while not consumed in a reaction, provide an alternative pathway that results in an increased reaction rate. This phenomenon is a cornerstone in numerous industrial processes, environmental management strategies, and the intricate biochemical reactions taking place within living organisms. Enzyme kinetics is a specialized branch of catalysis that delves into how biological catalysts, or enzymes, facilitate biochemical reactions—often with remarkable specificity and efficiency.

Catalysts function by offering an alternative reaction pathway with a lower activation energy, thereby increasing the number of molecular collisions with sufficient energy to reach the transition state. The presence of a catalyst modifies the potential energy landscape of a reaction without altering the reactants or products. This can be visually appreciated by contrasting the energy profile of an uncatalyzed reaction with that of a catalyzed one.

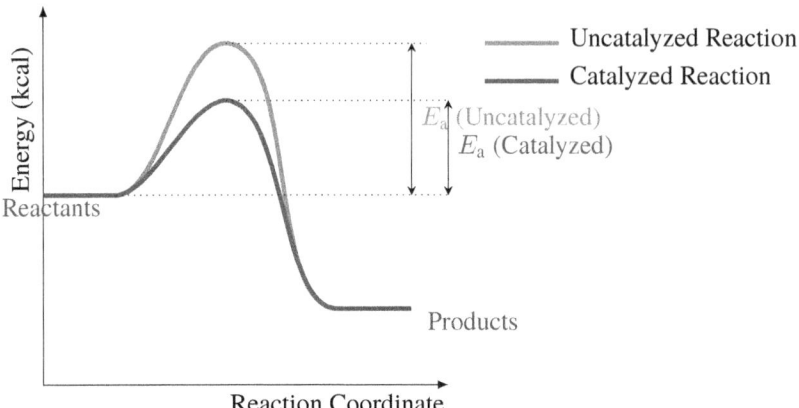

The lower activation energy in the catalyzed reaction (dashed line) reduces the energy barrier, thereby increasing the reaction rate.

Catalysis can be broadly categorized into homogeneous, heterogeneous, and biocatalysis, each distinguished by the phase of the catalyst relative to the reactants.

- **Homogeneous Catalysis:** Occurs when the catalyst and reactants are in the same phase, typically liquid. An example includes acid-catalyzed esterification reactions where sulfuric acid (H_2SO_4) catalyzes the formation of esters from alcohols and carboxylic acids.

- **Heterogeneous Catalysis:** Characterized by the catalyst existing in a different phase than the reactants, often occurring on the surface of a solid catalyst interacting with gaseous or liquid reactants. Industrial examples include the Haber process for ammonia synthesis, leveraging iron catalysts, and catalytic converters in vehicles that use platinum and palladium to convert harmful gases into less toxic emissions.

- **Biocatalysis (Enzyme Catalysis):** Involves biological molecules (enzymes) catalyzing biochemical reactions. Enzymes provide excellent selectivity and operate under mild conditions, exemplified by the catalysis of metabolic pathways in organisms.

5.5. CATALYSIS AND ENZYME KINETICS

Enzymes are sophisticated protein molecules that accelerate biochemical reactions with astounding precision and efficiency. They exhibit exceptional specificity for their substrates—the molecules upon which enzymes act—and function under physiological conditions of pH and temperature. Enzyme kinetics investigates the rate of enzymatic reactions, providing insights into enzyme activity, substrate affinity, and metabolic control.

A landmark model in enzyme kinetics is the Michaelis-Menten model, which describes the rate of enzyme-catalyzed reactions as a function of substrate concentration. This model simplifies a complex biochemical process into a sequence of steps, with the reaction proceeding through the formation of an enzyme-substrate complex (ES) that subsequently breaks down to release product and free enzyme.

The general enzymatic reaction follows:

$$E + S \underset{k_{-1}}{\overset{k_1}{\rightleftharpoons}} ES \xrightarrow{k_2} E + P$$

Under the steady-state approximation, the rate of change of the enzyme-substrate complex concentration $[ES]$ is assumed to be nearly zero. The resulting rate equation is given by:

$$v_0 = \frac{V_{\max}[S]}{K_m + [S]}$$

where: - v_0 is the initial reaction velocity. - $V_{\max} = k_2[E]_0$ is the maximum velocity at saturating substrate concentrations. - $[S]$ is the substrate concentration. - $K_m = (k_{-1} + k_2)/k_1$ is the Michaelis constant, indicative of the substrate concentration at half-maximal velocity, reflecting enzyme affinity for the substrate.

For practicality in deriving kinetic parameters, the Lineweaver-Burk plot, or double-reciprocal plot, is utilized to linearize the Michaelis-Menten equation:

$$\frac{1}{v_0} = \frac{K_m}{V_{\max}} \frac{1}{[S]} + \frac{1}{V_{\max}}$$

This equation is linearized to the form $y = mx + c$, allowing for easy determination of K_m and $V_{\max}$ from the slope and intercept, respectively.

Enzyme activity is highly sensitive to several factors:

- **Temperature:** Enzymes exhibit optimal activity within a specific temperature range. Elevated temperatures enhance reaction velocity but can lead to enzyme denaturation if excessively high.

- **pH:** Each enzyme operates best at an optimal pH reflective of its natural environment. Deviations from this pH can diminish enzyme activity by altering the enzyme's structure or the ionization state of its active site.

- **Inhibitors:** Chemicals that reduce enzyme activity by binding to the enzyme. They are categorized into competitive, non-competitive, and uncompetitive inhibitors, each affecting enzyme kinetics distinctively.

Understanding enzyme inhibition is essential in drug design and regulating metabolic pathways. Enzyme inhibitors act by reducing enzyme activity through various mechanisms:

- **Competitive Inhibition:** Occurs when the inhibitor competes with the substrate for the enzyme's active site. This raises the apparent K_m without affecting $V_{\max}$, as inhibition can be overcome by increasing substrate concentration.

- **Non-Competitive Inhibition:** The inhibitor binds to an allosteric site, undeterred by the presence of the substrate. It lowers $V_{\max}$ without affecting K_m, indicating inhibition cannot be reversed by excess substrate.

- **Uncompetitive Inhibition:** The inhibitor binds to the enzyme-substrate complex, decreasing both K_m and $V_{\max}$, with inhibition augmented by substrate concentration.

Catalysis and enzyme kinetics are integral to fields like industrial chemistry, pharmaceuticals, and biotechnology:

- **Pharmaceuticals:** Enzyme inhibitors form the basis of numerous drugs, including antibiotics, antidepressants, and protease inhibitors used in treating infections and conditions such as HIV/AIDS. These inhibitors modulate enzyme pathways, tailored to interrupt specific biochemical activities while minimizing side effects.

5.5. CATALYSIS AND ENZYME KINETICS

- **Industrial Processes:** Catalysts are essential in the petrochemical industry for refining crude oil, in polymer production for manufacturing plastics, and in green chemistry for developing sustainable processes that minimize waste.

- **Biotechnology and Agriculture:** Enzymes are exploited in agritech for developing eco-friendly pest control and crop enhancement solutions. Enzyme kinetics further aids in improving fermentation efficiency in alcohol and dairy production.

Advancements in enzyme engineering, such as directed evolution, enable the development of enzymes with tailored characteristics, expanding their applicability in harsh industrial conditions.

Ongoing research seeks to enhance understanding and control over catalytic processes, with several promising areas:

- **Biocatalyst Design:** Engineered enzymes and synthetic biomolecules are being developed for specific, high-efficiency catalysis in non-natural environments. Rational design and directed evolution contribute to creating catalysts with improved stability and novel functionalities.

- **Sustainable Catalysis:** Emphasis on developing catalysts that support green chemistry principles, reducing energy consumption and production of harmful byproducts, remains a key priority.

- **Integrative Kinetic Modeling:** Employing computational techniques to simulate complex catalytic processes, enabling predictive modeling of kinetics under diverse operational conditions.

The study of catalysis and enzyme kinetics is a dynamic and multifaceted domain critical for advancing scientific and engineering endeavors. Whether handling large-scale chemical production or understanding minute biochemical interactions within organisms, the principles of catalysis continue to inspire and direct innovative solutions to modern challenges. With the integration of cutting-edge technologies and interdisciplinary research approaches, the versatility of catalysis will remain a pivotal element in addressing global needs and driving chemical science forward.

5.6 Transition State Theory and Reaction Dynamics

Transition state theory (TST) and reaction dynamics are cornerstones of theoretical chemistry, providing profound insights into how chemical reactions proceed at the molecular level. TST offers a framework for understanding the critical configurations—transition states—through which reactants pass on their way to becoming products. Reaction dynamics complements this by studying the motion and interactions of atoms and molecules during the entire course of the reaction.

Transition state theory was developed in the 1930s by Henry Eyring, Michael Polanyi, and others. It fundamentally posits that chemical reactions occur when reactant molecules pass through a high-energy transition state. This state, also known as the activated complex, represents a point of maximum energy on the reaction coordinate, which is a conceptual pathway that depicts the progression of a reaction.

The transition state theory establishes that the rate of a reaction is determined by the concentration of the transition state and the frequency with which it converts to products. The rate constant (k) for a reaction is given by:

$$k = \frac{k_B T}{h} e^{-\Delta G^\ddagger / (RT)}$$

where k_B is the Boltzmann constant, h is Planck's constant, T is the temperature in Kelvin, and $\Delta G^\ddagger$ is the Gibbs free energy of activation. This expression eloquently links macroscopic reaction rates with molecular properties, providing a powerful predictive capability.

The concept of a potential energy surface (PES) is central to TST. A PES is a multidimensional surface representing the potential energy of a system as the positions of its nuclei change. In the simplest case, for a reaction involving two atoms or molecules, the PES can be visualized as a three-dimensional plot with reaction coordinates on the axes and potential energy on the vertical.

5.6. TRANSITION STATE THEORY AND REACTION DYNAMICS

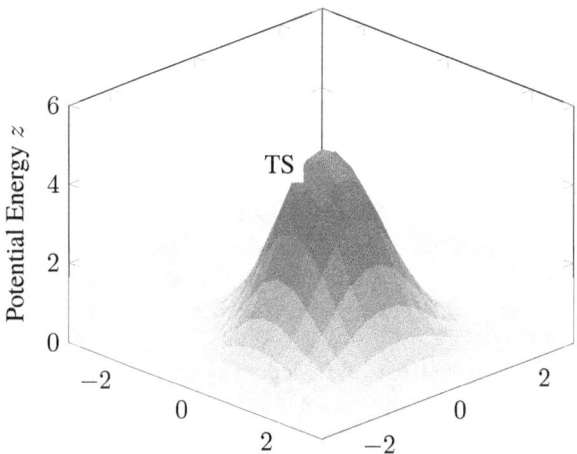

Reaction Coordinate x — Reaction Coordinate y

The highest point of the energy ridge on the PES corresponds to the transition state. Molecules must reach this energy checkpoint to transform from reactants to products. The topology of the PES influences the possible kinetic pathways and highlights regions where energy barriers hinder progress.

For most reactions, the transition state represents a fleeting, high-energy structure between reactants and products. Consider a simple reaction:

$$A + B \longrightarrow C + D$$

The energy profile typically includes:

- **Reactants:** Initial low-energy state.
- **Transition State:** Peak of the energy barrier that molecules must overcome.
- **Products:** Usually lower in energy than the transition state, often lower than reactants too, if the reaction is exothermic.

While transition state theory provides a statistical picture of how reactions proceed, reaction dynamics delves deeper into the real-time motion and interactions of molecules. This approach considers the forces and trajectories at play as molecules evolve from reactants through the transition state to become products.

Reaction dynamics utilize both experimental techniques and theoretical models to explore reactions:

- **Molecular Beam Experiments:** Techniques like crossed molecular beams allow physicists and chemists to investigate collisions between reactant molecules under controlled conditions. Observing scattering angles and energy distributions offers insights into reaction pathways.

- **Time-Resolved Spectroscopy:** Methods such as femtosecond and picosecond laser spectroscopy capture the rapid changes in a system's energy levels as the reaction progresses. This elucidates the timing and formation of intermediates and transition states.

- **Molecular Dynamics Simulations:** Computational methods model the behavior of molecules during reactions, predicting trajectories and energies that unfold at atomic levels. By simulating thousands of possible paths, this approach helps identify preferred routes and potential bottlenecks.

The principles underlying transition state theory and reaction dynamics are applied in multiple scientific fields. A few notable examples include:

- **Chemical Synthesis:** Predicting reaction kinetics and pathways is essential in organic and inorganic synthesis optimization. By understanding potential energy surfaces, chemists can manipulate conditions to favor desired products and minimize side reactions.

- **Enzyme Functionality:** In biochemistry, enzymes work by stabilizing transition states, substantially lowering the activation energy. This stabilization is quantified through reaction dynamics studies, revealing how enzymes orchestrate biological transformations with high specificity.

- **Catalysis:** Industrial catalysts function by offering pathways with modified transition states. Understanding these modifications through TST can lead to improved catalyst design and more efficient chemical processes.

5.6. TRANSITION STATE THEORY AND REACTION DYNAMICS

Transition states play a pivotal role in determining reaction rates and selectivity. They embody the critical point where potential energy, molecular geometry, and electronic structure are finely balanced. This state is not only a theoretical construct; modern laser and spectroscopic techniques have made it possible to observe transition-state structures albeit indirectly. They present fleeting glimpses of molecules poised for transformation, cementing their importance in both theoretical and practical chemistry.

Despite the remarkable advances enabled by TST and reaction dynamics, challenges persist:

- **Complex Systems:** Real-world reactions, especially those in biological or condensed matter environments, involve numerous atoms and variables. Accurate modeling of such systems requires sophisticated computational resources and methodologies to account for all interactions.

- **Non-Classical Pathways:** Reactions involving tunneling and other quantum phenomena do not conform neatly to classical transition state theory. Incorporating quantum principles into dynamics simulations continues to challenge researchers.

- **Interdisciplinary Integration:** Ongoing advancements benefit from cross-disciplinary collaboration, uniting quantum physics, computational science, and chemical engineering to tackle complex reaction dynamics.

Looking forward, the continuous development of computational power and analytical techniques promises deeper insights. Exploring new quantum algorithms and enhanced spectroscopic resolutions can reveal more about the nuances of real-world chemical reactions. Future work in TST and reaction dynamics will strive to integrate these progresses, unraveling the intricate dance of atoms and electrons that underpins chemical change.

Transition state theory and reaction dynamics form the core of understanding how and why chemical reactions occur. By elucidating the high-energy states and detailed motions within reactions, chemists and scientists leverage this understanding to innovate in synthesis, catalysis, and materials science, driving the field of chemistry towards new frontiers.

5.7 Collision Theory and Molecular Dynamics Simulations

Collision theory and molecular dynamics simulations play a crucial role in the study of chemical kinetics, elucidating the microscopic interactions that underpin reaction processes. Collision theory offers a simplified model explaining the occurrence of chemical reactions, attributing their progress to the frequency and energy of collisions between reactive molecules. In contrast, molecular dynamics simulations provide a computational approach to dynamically model the motions and interactions of molecules, offering detailed insights into the transition of atoms from reactants to products.

Collision theory posits that for a reaction to occur, reactant molecules must collide with sufficient energy and proper orientation. The theory emphasizes three critical factors:

- **Collision Frequency:** The number of collisions occurring per unit time. An increase in reactant concentration results in more collisions, enhancing the likelihood of reaction.

- **Activation Energy:** Despite the collision frequency, reactant molecules must possess a minimum energy, known as the activation energy (E_a), to initiate a reaction. This energy overcomes the potential energy barrier to reach the product state.

- **Proper Orientation:** For successful collisions, molecules must orient in a specific manner. Only collisions in which the reactive moieties are favorably aligned lead to chemical transformation.

The rate of a bimolecular reaction can be expressed as:

$$\text{Rate} = Z[A][B]e^{-E_a/(RT)}$$

where Z factors in collision frequency concerning molecular orientation.

The success of a collision depends crucially on both the kinetic energy of colliding molecules and their orientation. This is visualized by the molecular energy distribution in a sample, typically depicted by the Maxwell-Boltzmann distribution curve.

5.7. COLLISION THEORY AND MOLECULAR DYNAMICS SIMULATIONS

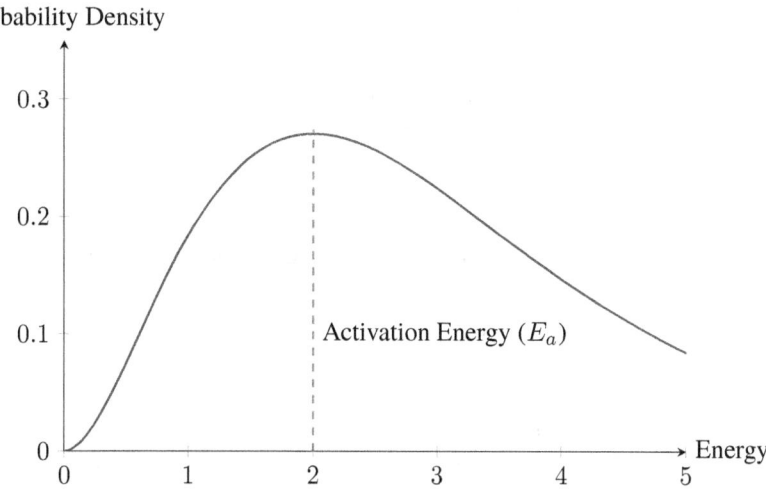

The graph shows that only some molecules have energy equal to or greater than the activation energy, which is necessary for a reaction. An increase in temperature leads to a broader distribution, indicating a higher fraction of molecules capable of successful collisions.

Proper orientation is also crucial. Not every collision, even among molecules with energy exceeding E_a, leads to a reaction. Specific molecular orientations are necessary for effective orbital overlap during bond breaking and formation.

While collision theory provides a heuristic understanding, the complexity of chemical reactions in actual systems requires a more detailed investigation. Molecular dynamics (MD) simulations offer a computational approach that captures the temporal evolution of molecular systems, incorporating Newtonian mechanics to simulate particle behavior.

- **Initial Setup:** Define initial atomic coordinates and velocities based on known structures and experimental data or theoretical predictions.

- **Force Calculation:** Use potential energy functions or force fields to determine interactions between atoms. Commonly used force fields include CHARMM, AMBER, and GROMOS, which model bond stretching, angle bending, torsional rotations, and non-bonded interactions like van der Waals forces and electrostatic interactions.

- **Time Propagation:** Use integration algorithms, such as the Verlet or leapfrog method, to iteratively update positions and velocities, advancing the system in small time steps (femtoseconds).
- **Analysis of Trajectories:** Examine trajectories to understand dynamic behavior, structural properties, conformational changes, and reaction pathways.

These simulations provide insights into microscopic mechanisms of reactions and the role of molecular motions, offering explanations for observed kinetic behaviors.

Molecular dynamics simulations have transformative potential across various scientific disciplines:

- **Chemical Kinetics:** MD simulations enable detailed exploration of reaction mechanisms, elucidating complex pathways and intermediate states inaccessible through experimental techniques alone. They have been extensively used to model solutions and gas-phase reactions, helping to understand solvent effects and collision dynamics.
- **Materials Science:** Investigating properties of nanomaterials and polymers at the molecular level reveals how atomic interactions influence macroscopic properties. Simulations facilitate tunneling phenomena studies, such as in semiconductors and insulators.
- **Biochemistry and Drug Design:** Understanding protein dynamics, ligand binding affinity, and enzyme catalysis through MD simulations supports drug discovery efforts. They predict binding sites, conformational changes upon ligand association, and allosteric effects, significantly reducing costs and time in drug development processes.
- **Environmental Chemistry:** Simulations clarify pollutant interactions with atmospheric molecules, aiding assessments of environmental impacts and the reactive behavior of pollutants under various atmospheric conditions.

Despite their valuable insights, MD simulations face several limitations:

- **Computational Demand:** Simulations of large biomolecules or complex systems demand considerable computational resources and time, requiring supercomputers or specialized hardware like GPUs.

5.7. COLLISION THEORY AND MOLECULAR DYNAMICS SIMULATIONS

- **Accuracy of Force Fields:** The validity of results heavily depends on the accuracy of force fields, which are approximations. Emerging methods, such as *ab initio* MD, integrate quantum mechanical calculations to improve precision, though at a higher computational cost.

- **Time Scale Limitations:** Typical MD simulations cover timescales from nanoseconds to microseconds, shorter than the milliseconds to seconds needed for slow biological processes or rare event sampling. Techniques like enhanced sampling methods, including metadynamics and umbrella sampling, are used to overcome these barriers.

As computational power continues to increase and algorithms advance, the scope and accuracy of MD simulations will expand, providing deeper, more nuanced insights into the molecular choreography that underpins chemical reactions.

Progress in collision theory and molecular dynamics will increasingly rely on hybrid approaches combining classical and quantum mechanical theories to tackle complex systems:

- **QM/MM Simulations:** Quantum mechanics/molecular mechanics (QM/MM) methods simulate regions of a system at quantum levels while treating the rest classically, synergizing detailed chemical insights with manageable computational demands.

- **Machine Learning and Artificial Intelligence:** Incorporating AI techniques in developing reactive force fields or guiding simulations through efficient sampling can accelerate research timelines, providing predictive models and real-time reaction pathway optimization.

Collision theory and molecular dynamics simulations are powerful methodologies for investigating reaction mechanisms. Together, they provide a comprehensive view of the connection between molecular structure, energy landscapes, and kinetic phenomena. As technologies and methodologies advance, their application will continue bridging gaps in theoretical and experimental chemistry, expanding our understanding of molecular behavior.

Chapter 6

Statistical Mechanics and Thermodynamic Models

Statistical mechanics bridges the microscopic interactions of individual particles and the macroscopic laws of thermodynamics, offering a comprehensive framework for understanding system behavior at an atomic level. This chapter delves into the probabilistic underpinnings of statistical mechanics, explaining how microstates and macrostates contribute to emergent thermodynamic properties. Through the exploration of partition functions and Boltzmann distributions, the link between molecular energies and observable phenomena is established. The discourse extends to the application of distinct thermodynamic models and ensembles, such as canonical and grand canonical, which aid in predicting system behavior under varied constraints. By integrating classical and quantum statistics, the chapter illuminates the profound connection between microscopic interactions and global thermodynamic trends.

6.1 Fundamentals of Statistical Mechanics

Statistical mechanics offers a comprehensive framework for understanding the macroscopic properties of systems based on the microscopic behaviors of their constituent particles. At its core, statistical mechanics rests on the definitions and distinctions between microstates and macrostates, which provide crucial insights into the probabilistic nature of physical systems.

A *microstate* refers to a specific configuration of a system on a microscopic level, encompassing the positions and momenta of all particles within the system. For a gas contained in a volume, each distinct allocation of particle velocities and positions corresponds to a distinct microstate. Contrarily, a *macrostate* denotes the macroscopic description of the system, characterized by measurable properties such as temperature, pressure, volume, or energy. A macrostate is essentially an aggregation of various microstates that conform to the observable parameters.

The key to statistical mechanics lies in the vast number of microstates that can correspond to a single macrostate. This multitude imbues systems with inherent probabilistic attributes, leading to an emergent behavior that is consistent with classical thermodynamics. The primary principle guiding such systems is the assumption of *equal a priori probabilities*, which posits that, in an isolated system at equilibrium, all accessible microstates are equally probable. This principle is pivotal in determining the probability distribution across different microstates.

To elucidate the role of probability, consider the mechanical model of an ideal gas. Each molecule in the gas can be in a state with specific position (x, y, z) and momentum (p_x, p_y, p_z). The phase space, a six-dimensional space defined by these components, dictates the behavior of a single molecule. For N non-interacting molecules, the phase space is consequently a $6N$-dimensional ensemble, representing all possible microstates.

The number of microstates Ω, often termed as the multiplicity of a macrostate, is a fundamental quantity in statistical mechanics. It serves as an indicator of the entropy S, according to Boltzmann's famous relation:

$$S = k_B \ln \Omega$$

where k_B is the Boltzmann constant. Entropy, a central concept in thermodynamics, measures the degree of disorder or randomness of a system. In statistical mechanics, it is given a precise definition related to the logarithm of the number of microstates, expressing an intrinsic connection between microscopic configurations and macroscopic disorder.

The concept of *ensemble* is integral to the probabilistic approach in statistical mechanics. An ensemble is a hypothetical collection of a very large number of

6.1. FUNDAMENTALS OF STATISTICAL MECHANICS

virtual copies of the system, considered in all possible microstates consistent with the macroscopic conditions. Three primary types of statistical ensembles are employed, depending on the specific constraints:

- **Microcanonical ensemble**: Represents an isolated system with fixed energy, volume, and particle number. It describes a system's behavior when it is not exchanging energy or matter with the environment.

- **Canonical ensemble**: Used for closed systems that can exchange energy with a heat bath at fixed temperature, but remain closed to particle exchanges. This ensemble fixes the number of particles, volume, and temperature.

- **Grand canonical ensemble**: Suited for systems that can exchange both energy and particles with a reservoir, thus having fixed temperature, volume, and chemical potential.

In each case, the ensemble facilitates the calculation of thermodynamic properties by considering the probability of each microstate. The ensemble's distribution function assigns probabilities to the microstates compatible with the constraints and thus determines the thermodynamics of the system.

The probability of finding the system in a particular state is expressed by the following relation in the canonical ensemble:

$$P_i = \frac{e^{-\beta E_i}}{Z}$$

where E_i is the energy of the i-th microstate, $\beta = \frac{1}{k_B T}$ is the inverse thermal energy, and Z is the partition function, defined as:

$$Z = \sum_i e^{-\beta E_i}$$

The partition function Z serves as a crucial tool, summing over all microstates and capturing the statistical properties of the system. It acts as a bridge between the microscopic states and macroscopic observables by linking various thermodynamic quantities through relations such as:

$$F = -k_B T \ln Z$$

where F is the Helmholtz free energy. The partition function also enables the calculation of average energy, entropy, and specific heat, among other quantities.

Consider an illustrative example involving a simple spin system. Let us examine a magnetic moment in a magnetic field, often referred to as the two-level system. In this scenario, the magnetic moment can assume one of two states: spin-up with energy $-\mu B$ or spin-down with energy μB, where μ is the magnetic moment and B is the magnetic field strength. The partition function for this system is expressed as:

$$Z = e^{\beta \mu B} + e^{-\beta \mu B}$$

From the partition function, we can deduce the average energy $\langle E \rangle$ and magnetization M:

$$\langle E \rangle = -\frac{\partial \ln Z}{\partial \beta}, \quad M = \frac{\mu}{k_B T} \tanh(\beta \mu B)$$

These results reflect the fundamental behavior of the two-level system, demonstrating the utility of statistical mechanics in predicting system characteristics based on microstates.

The fundamentals of statistical mechanics encompass the concepts of microstates, macrostates, and probability distributions, constructing a robust framework for analyzing thermodynamic systems. By probabilistically mediating the transition from the microscopic to the macroscopic, statistical mechanics elucidates the intrinsic nature of entropy and enables the calculation of thermodynamic properties across diverse physical contexts.

6.2 Boltzmann Distribution and Molecular Energies

The Boltzmann distribution is a cornerstone of statistical mechanics, articulating the distribution of energies among molecules in a system at thermal equilibrium. It offers profound insights into energy partitioning, ranging from gases and liquids to the vibrational modes of solids. The Boltzmann distribution helps in understanding molecular energies and elucidates essential thermodynamic behaviors.

6.2. BOLTZMANN DISTRIBUTION AND MOLECULAR ENERGIES

In a system governed by thermal fluctuations, the probability P_i of finding a molecule in a particular energy state E_i is influenced by the temperature T of the system and is expressed as:

$$P_i = \frac{e^{-\beta E_i}}{Z}$$

where $\beta = \frac{1}{k_B T}$ is the inverse thermal energy with k_B being the Boltzmann constant, and Z is the partition function defined as:

$$Z = \sum_i e^{-\beta E_i}$$

This expression indicates that at equilibrium, molecules occupy higher-energy states with an exponentially smaller probability compared to lower-energy states, emphasizing the role of temperature in distributing molecular energies.

The Boltzmann distribution formula is derived under the assumption of a system composed of non-interacting particles at thermal equilibrium. This derivation is often approached by maximizing entropy subject to constraints of a fixed number of particles and a fixed average energy, leveraging techniques from calculus of variations and Lagrange multipliers.

Consider a mixture of gas molecules, each capable of occupying distinct energy levels. At a given temperature, the system's thermal energy allows it to explore these states. In this equilibrium, lower-energy states are more populated, adhering to the trade-off between minimizing the system's energy and maximizing its entropy.

An illustrative manifestation of the Boltzmann distribution is the Maxwell-Boltzmann distribution, which specifies the distribution of kinetic energies among molecules in an ideal gas. The kinetic energy E of a molecule with velocity v is given by:

$$E = \frac{1}{2}mv^2$$

The probability density function for the energy distribution is:

$$f(E) = \left(\frac{2\pi}{k_B^3}\right)^{-1/2} \left(\frac{1}{m}\right)^{3/2} E^{1/2} e^{-\frac{E}{k_B T}}$$

This distribution highlights that the most probable speed is not at the lowest energy level, an implication of the multiplicity factor $E^{1/2}$ which favors higher speeds due to the increased number of states of higher energy.

Exploring applications beyond gases, consider the distribution of molecular energies in vibrational and rotational modes within polyatomic molecules. These discrete quantized energy levels follow Boltzmann statistics. For example, in a solid, atoms vibrate around their equilibrium positions, and the distribution of these vibrational energy levels follows:

$$P_n = \frac{e^{-n\hbar\omega/k_B T}}{Z}$$

where $\hbar$ is the reduced Planck's constant and ω is the angular frequency of the oscillator mode. This expression is crucial for understanding phenomena such as specific heat capacities of solids as temperature changes.

Moreover, the Boltzmann distribution adeptly explains rates of chemical reactions. The Arrhenius equation, expressing reaction rates as a function of temperature, incorporates the concept that molecules must overcome an energy barrier (activation energy E_a) to react:

$$k = Ae^{-E_a/k_B T}$$

Here, the exponential term, a direct consequence of the Boltzmann distribution, governs how temperature facilitates the transition over this energy barrier, explaining why reactions speed up with increasing temperatures.

A fascinating case is the distribution of energy states in semiconductors, where electrons populate conduction bands depending on thermal availability. The Fermi-Dirac statistics extend Boltzmann's perspective by accounting for Pauli's exclusion principle, yet under conditions where $E \gg k_B T$, Fermi-Dirac statistics simplifies to the Boltzmann form, confirming its versatility in predicting carrier distribution in intrinsic and extrinsic semiconductors at varied temperatures.

Besides, the Boltzmann distribution applies to the field of astronomical sciences, where it helps describe the distribution of stellar velocities, signaling how gravitational and thermal dynamics dictate star motions within galaxies. The Galactic rotation curves further reflect the distribution governed by such equilibrium concepts, intricately linked to dark matter distribution models.

Analyzing deviations from the Boltzmann distribution illuminates non-equilibrium phenomena. In plasmas, high-energy particles follow forms of distribution that slightly deviate, leading to new understandings in hot astrophysical gases or controlled fusion research.

In experimental terms, spectroscopy depends on discerning energy distributions in molecular systems. Techniques such as infrared, Raman, or nuclear magnetic resonance (NMR) spectroscopy observe transitions between these energy levels, providing direct measurements of the population of states described by Boltzmann statistics. For proteins, the Boltzmann-weighted ensemble average predicts observed properties in solution, guiding how macromolecular structures relate to thermal fluctuations.

To appreciate the breadth of Boltzmann distribution's utility, one must evaluate its contribution to statistical thermodynamics, where microscopic states account for macroscopic observables. The equilibrium constant of chemical reactions emerges from Boltzmann factors, dictating how reactants and products balance at a given temperature.

Entropy production and dissipation, within irreversible processes, further extends Boltzmann's concepts into non-equilibrium realms, prompting sophisticated theories like fluctuation theorems that encapsulate deviations from equilibrium based on probability distributions.

The Boltzmann distribution is not merely a mathematical artifact—it is a profound descriptive tool embodying the very nature of thermodynamic systems. By unveiling the statistical foundation of molecular energies, it interlinks diverse phenomena spanning chemical kinetics, materials science, astrophysics, and beyond, remaining an indispensable framework as we delve into complex systems' dynamic behaviors in both equilibrium and emerging non-equilibrium contexts.

6.3 Partition Functions and their Applications

Partition functions serve as the foundation of statistical mechanics, providing a mathematical vehicle to connect the microscopic states of a system with its macroscopic thermodynamic properties. In essence, partition functions encode the statistical information of all possible configurations a system can attain, rendering them indispensable for analyzing physical systems at equilibrium.

The partition function, typically denoted as Z, is a sum over all possible microstates, where each microstate i has an energy E_i, represented mathematically for a canonical ensemble as:

$$Z = \sum_i e^{-\beta E_i}$$

Here, $\beta = \frac{1}{k_B T}$, with k_B being the Boltzmann constant and T the absolute temperature. This fundamental construct acts as a normalization factor ensuring the probability distribution over states sums to one, central to determining probabilities of various states and the ensuing thermodynamic properties.

Thermodynamic Properties Derived from Partition Functions

The partition function's utility extends far beyond merely cataloging states; it serves as the gateway to calculating vital thermodynamic quantities. The Helmholtz free energy F is directly related to the partition function as:

$$F = -k_B T \ln Z$$

This relationship allows us to derive other thermodynamic properties such as internal energy U, entropy S, and specific heat C_V:

$$U = -\frac{\partial \ln Z}{\partial \beta}, \quad S = k_B (\ln Z + \beta U), \quad C_V = \frac{\partial U}{\partial T}$$

Each of these properties is rooted in the microstate distribution, thereby translating ensemble statistics to observable macroscopic quantities.

Partition Functions for Different Ensembles

Though the canonical partition function is quintessential, it's crucial to understand that partition functions vary across different ensembles, each catering to specific constraints. The microcanonical partition function is concerned with systems of fixed energy, particle number, and volume, essentially counting the number of accessible microstates at a given energy, represented as:

6.3. PARTITION FUNCTIONS AND THEIR APPLICATIONS

$$\Omega(E, V, N) = \sum_{\{i|E_i=E\}} 1$$

In contrast, the grand canonical partition function, accommodating systems exchanging both energy and particles with a reservoir, incorporates both the chemical potential μ and is defined as:

$$\Xi = \sum_{N=0}^{\infty} e^{\beta \mu N} Z_N$$

where Z_N is the canonical partition function for the system with N particles.

Applications in Ideal and Non-Ideal Gases

The partition function for an ideal gas is particularly illustrative of its practical applications. The canonical partition function Z for a single particle in an ideal gas is derived by integrating over all possible states:

$$Z_1 = \frac{1}{h^3} \int e^{-\beta\left(\frac{p^2}{2m}+U\right)} d^3r\, d^3p$$

where h is Planck's constant, U is the potential energy, and the integral runs over all space and momentum. For a non-interacting multi-particle system, the total partition function becomes:

$$Z = \frac{Z_1^N}{N!}$$

The factorial accounts for the indistinguishability of particles, a critical adjustment in quantum statistics. This derivation leads to expressions for the pressure P and entropy S that readily yield the ideal gas law.

Beyond ideal gases, the partition function is instrumental in analyzing systems where interactions cannot be overlooked. For example, in Van der Waals gases, corrections to the ideal gas law account for molecular volume and intermolecular attractions, leading to modified partition functions that capture these effects.

Quantum Mechanical Systems and Partition Functions

In quantum mechanical systems, the degeneracy of energy states underscores the pivotal role of partition functions. For instance, consider the harmonic oscillator, where energy levels are quantized as:

$$E_n = \hbar\omega \left(n + \frac{1}{2}\right)$$

The corresponding partition function is expressed as an infinite geometric series:

$$Z = \sum_{n=0}^{\infty} e^{-\beta\hbar\omega(n+\frac{1}{2})} = \frac{e^{-\frac{1}{2}\beta\hbar\omega}}{1 - e^{-\beta\hbar\omega}}$$

From this, one derives the average energy, entropy, and heat capacity, essential for exploring quantum phenomena such as zero-point energy and anharmonic effects at low temperatures.

Partition Functions in Chemical Reactions

The significance of partition functions extends robustly into the realm of chemical reactions. They pave the way for calculating equilibrium constants, essential for understanding reaction dynamics. The equilibrium constant K is related to the change in free energy, which connects through partition functions:

$$K = \exp\left(-\frac{\Delta G^\circ}{k_B T}\right) = \prod_i \left(\frac{q_i^B}{q_i^A}\right)^{\nu_i}$$

where q_i^B and q_i^A denote partition functions per molecule, and ν_i the stoichiometric coefficients.

Biological and Astrochemical Applications

In biological systems, partition functions help clarify enzyme activity and molecular signaling pathways. For example, they facilitate the understanding of protein folding landscapes, where the stability of folded versus unfolded states at varying temperatures is evaluable through partition function calculations, providing insights into thermodynamic stability.

In astrochemistry, partition functions play a critical role in understanding molecular abundances in diverse environments, from interstellar media to planetary atmospheres. Rotational and vibrational partition functions allow prediction of spectral line intensities, crucial for inferring physical conditions and chemical compositions of distant astronomical entities.

Extensions Beyond Equilibrium - Nonequilibrium Statistical Mechanics

Though traditional partition functions assume equilibrium, extensions to nonequilibrium systems are profound. In systems far from equilibrium, like driven diffusive systems or biological networks, generalized partition functions accommodate steady states, occasionally described through nonequilibrium work relations or stochastic thermodynamics, which probe systems' behavior under continuous external influences.

The partition function is an infinitely valuable tool summarizing the statistical landscape of a physical system and making possible the translation from microscopic dynamics to macroscopic observables across physical, chemical, and biological disciplines. Through its varied applications, it exemplifies the elegance and power inherent in the statistical mechanics framework, continuously driving new insights in theoretical and applied fields.

6.4 Classical and Quantum Statistics

The discipline of statistical mechanics bifurcates into classical and quantum domains, each equipped with distinct principles and methodologies to describe the behavior of particles. While classical statistics apply to macroscopic realms where quantum effects are negligible, quantum statistics become crucial at microscopic scales where phenomena such as indistinguishability and quantum symmetries dominate. Understanding these frameworks provides a comprehensive view of physical systems, revealing how emergent macroscopic properties stem from underlying microscopic principles.

Classical Statistics: Maxwell-Boltzmann Distribution

In classical statistics, particles are distinguishable and adhere to the Maxwell-Boltzmann distribution. This framework is valid when the de Broglie wavelength of particles is small compared to the average separation distance, typically applying to high-temperature and low-density conditions.

For a gas of N non-interacting classical particles, the fraction of particles occupying a state i with energy E_i is given by:

$$f_i = \frac{g_i e^{-\beta E_i}}{Z}$$

where g_i denotes the degeneracy of the state and Z is the partition function. The Maxwell-Boltzmann distribution effectively predicts properties like pressure, temperature, and internal energy for ideal gases, and by extension, fundamental laws such as the equipartition theorem, which asserts that each degree of freedom contributes $\frac{1}{2} k_B T$ to the system's average energy.

Transcending individual predictions, classical statistics elucidate transport phenomena. For instance, the Maxwell-Boltzmann distribution provides the basis for deriving mean free path, diffusion coefficients, and thermal conductivity in gases through the Chapman-Enskog expansion, ensuring coherence with macroscopic observations such as viscosity and heat conduction laws.

Limitations of Classical Statistics

Despite its successes, classical statistics exhibit limitations notably under conditions where quantum effects become significant. When particles exhibit wave-like behavior, as manifest in systems at cryogenic temperatures or of high density, indistinguishability and quantization of energy levels necessitate a quantum mechanical treatment. This is where classical assumptions falter, making quantum statistics indispensable.

Quantum Statistics: Fermi-Dirac and Bose-Einstein Distributions

Quantum statistics bifurcates into two distinct formulations: Fermi-Dirac statistics for fermions and Bose-Einstein statistics for bosons, reflecting the fundamental distinctions between these quantum particles.

- **Fermi-Dirac Statistics**: Fermions, such as electrons, protons, and neutrons, conform to the Pauli exclusion principle, where no two fermions can occupy the same quantum state simultaneously. The distribution

6.4. CLASSICAL AND QUANTUM STATISTICS

function for fermionic particles is given by:

$$f_i = \frac{1}{e^{\beta(E_i-\mu)}+1}$$

Here, μ is the chemical potential, and this distribution implies that at zero temperature, fermions completely fill up energy states up to the Fermi energy E_F. Such behavior characterizes systems like electron gases in metals, white dwarf stars, and atomic nuclei, where Pauli repulsion stabilizes matter against gravitational collapse or thermodynamic equilibrium.

Fermi-Dirac statistics underpin solid-state physics, aiding in the characterization of electronic properties of materials, encompassing concepts such as band structure, density of states, and electrical conductivity. Furthermore, they explicate key phenomena like the Quantum Hall Effect and superconductivity, where quantum coherence at low temperatures leads to macroscopic quantum phenomena.

- **Bose-Einstein Statistics**: Conversely, bosons including photons, helium-4 nuclei, and gluons, do not adhere to the exclusion principle; multiple bosons can occupy the same state, leading to entirely different statistical behavior. The Bose-Einstein distribution is formulated as:

$$f_i = \frac{1}{e^{\beta(E_i-\mu)}-1}$$

Bose-Einstein condensation epitomizes this behavior, where below a critical temperature, a macroscopic number of bosons occupy the lowest energy state, forming a condensate. This phenomenon is observed in superfluid helium and dilute gas Bose-Einstein condensates, observable in specialized laboratory settings.

Bose-Einstein statistics also describe the radiation field, laying the groundwork for black-body radiation theory, conceptualized through Planck's law, and facilitating developments like the laser, characterized by coherent photons, all sharing the same quantum state.

Interrelationship and Applications in Physics

The divergence between classical and quantum statistics converges near the classical limit. When examining high-temperature or sparse-particle scenarios, both Fermi-Dirac and Bose-Einstein distributions reduce to the classical

Maxwell-Boltzmann statistics. Such correspondence validates classical methods in appropriate regimes while highlighting quantum statistics' necessity in others.

Thermodynamics of degenerate quantum gases, a pristine exemplification, unfolds how temperature, pressure, and entropy behave under quantum restrictions. For fermions, degeneracy pressure governs dynamics uncaptured by classical treatments, elucidating behavior in astrophysical objects like neutron stars. Whereas, for bosons, transition dynamics provide insights into coherence levels of quantum fields.

Exploring the subtle integrals over these distributions reveals temperature-dependent phenomena, such as heat capacity and magnetic susceptibility variances in metals and magnetic materials, advancing the understanding of temperature-induced phase transitions.

Another remarkable application in polyatomic gases, where quantum effects in rotational and vibrational motions of molecules alter entropy and energy distributions, institutes corrections pivotal for precise modeling in atmospheric and chemical processes.

Statistical Insights into Quantum Information and Computation

Quantum statistics are also pivotal in quantum information science. Quantum entanglement, a direct offspring of indistinguishability and superposition principles, strengthens quantum computation, encryption, and teleportation theories, redefining transmission and processing of information paradigmatically distinct from classical frameworks.

Entropic measures extending from quantum states predict correlations and coherence within quantum systems, redressing boundaries of information theory and equilibrium thermodynamics towards controlling decoherence and error in quantum computing designs.

In summary, classical and quantum statistics collectively form the backbone of understanding a plethora of physical phenomena, establishing a robust platform upon which modern theoretical and applied physics flourish. As we probe deeper into the quantum realm, burgeoning fields like ultracold atom physics, quantum field theories, and condensed matter systems continually reaffirm the relevance and depth inherent in the statistics of classical and quantum domains, each reinforcing the other within their respective limits and uniting to provide a comprehensive picture of nature's underlying symmetries and complexities.

6.5 Thermodynamic Ensembles: Canonical, Microcanonical, and Grand Canonical

In statistical mechanics, ensembles are theoretical constructs that represent vast collections of possible microstates of a system, providing a powerful framework for deriving macroscopic thermodynamic properties from microscopic principles. These ensembles offer different ways to comprehend systems under various constraints, each uniquely suited to a particular set of conditions. The primary ensembles central to statistical mechanics are the microcanonical, canonical, and grand canonical ensembles, which differ in how they treat energy, particle number, and volume.

- **Microcanonical Ensemble**: The microcanonical ensemble is the simplest of the three, suited to isolated systems that do not exchange energy or matter with their surroundings. This ensemble is characterized by fixed total energy E, number of particles N, and volume V. The microcanonical ensemble describes a system at equilibrium with a uniform probability distribution over accessible microstates, assuming that all microstates compatible with these constraints are equally probable.

  The microcanonical partition function, often denoted as $\Omega(E, V, N)$ or the density of states, counts the number of microstates with a specific energy:

  $$\Omega(E, V, N) = \sum_{\{i | E_i = E\}} 1$$

  This enumeration reflects the system's total phase space volume and directly associates with the system's entropy S, as given by:

  $$S = k_B \ln \Omega$$

  The microcanonical ensemble is instrumental in building the foundation of thermodynamics by linking statistical definitions of entropy to its macroscopic counterpart. Moreover, it underpins the derivation of fundamental laws such as the second law of thermodynamics, delineating how isolated systems naturally evolve towards macrostates with maximum entropy.

- **Canonical Ensemble**: The canonical ensemble extends beyond the microcanonical by introducing temperature as a controlling variable. This ensemble applies to closed systems that can exchange energy with a thermal reservoir and, as such, are maintained at a constant temperature T. The number of particles N and volume V remain constant, but the energy E of the system fluctuates due to exchanges with the reservoir, allowing exploration across all energy states with probable weightings.

Within the canonical ensemble, the probability P_i of the system residing in a state i with energy E_i is given by:

$$P_i = \frac{e^{-\beta E_i}}{Z}$$

Here, Z represents the canonical partition function, central to the canonical ensemble, defined as:

$$Z = \sum_i e^{-\beta E_i}$$

This partition function bridges macroscopic thermodynamic observables with microscopic state probabilities, facilitating calculations of equilibrium properties such as free energy, internal energy, and specific heat. The Helmholtz free energy F, derived from the canonical partition function, is given by:

$$F = -k_B T \ln Z$$

This relation underscores the impact of state distribution on thermodynamic stability and spontaneous processes.

In practical terms, the canonical ensemble is vital in predicting system behaviors subject to thermal conditions, such as chemical reactions proceeding under controlled temperatures and bulk properties of materials under thermal equilibrium.

- **Grand Canonical Ensemble**: The grand canonical ensemble further generalizes the statistical description by permitting both energy and particle number fluctuations. Applicable to open systems in contact with both a thermal and particle reservoir, this ensemble fixes temperature T, volume V, and chemical potential μ.

The grand canonical partition function Ξ is computed as:

$$\Xi = \sum_{N=0}^{\infty} e^{\beta \mu N} Z_N$$

where Z_N is the canonical partition function for the system with N particles. It corresponds to systems where particle exchange is significant, and Ξ serves as the precursor to calculating the grand potential Φ_G:

$$\Phi_G = -k_B T \ln \Xi$$

From the grand canonical ensemble, fundamental thermodynamic variables such as particle number, pressure, and susceptibility emerge; it addresses phenomena like adsorption, evaporation, and reactions in gas phases.

- **Applications and Illustrations of Ensembles**:

- **Microcanonical Ensemble**: Consider the example of a closed adiabatic container filled with a gas. The microcanonical ensemble aptly describes the balance between kinetic and potential energies at different positions within the gas molecules in a perfectly insulated container. Such conditions echo cosmic scenarios like the interiors of stars, where energy constraints dominate dynamics, or closed-system calorimetric experiments probing reaction heat without environmental interactions.

- **Canonical Ensemble**: The canonical ensemble proves crucial for studying liquid-vapor transitions, such as the phase behavior of water under varied temperature and pressure. Precise temperature controls permit phase diagrams construction for numerous chemicals, highlighting critical points and coexistence regions pivotal for scientific and industrial applications like distillation.

   Another significant application is molecular dynamics simulations, where canonical ensembles model interactions under specified thermal conditions, translating microscopic motion into macroscopic observables such as viscosity, diffusivity, and equation of state parameters.

- **Grand Canonical Ensemble**: A quintessential example relates to semiconductors where doping adjusts electron and hole densities, influencing device conductivity. At constant temperature and potential, the grand canonical ensemble elucidates intrinsic and extrinsic carrier concentrations, impacting transistor performance and photovoltaic efficiency.

   Further, the grand canonical ensemble plays a pivotal role in adsorptive behavior on surfaces, studying how gases adhere to solids, pivotal for catalysis, sensor development, and environmental pollution filtration.

- **Ensemble Equivalence and Limitations**: It is noteworthy that under specific conditions, notably in the thermodynamic limit where $N \to \infty$ and volume $V \to \infty$ with N/V constant, the distinct ensembles converge in their predictions, affirming ensemble equivalence. However, exceptions arise in finite systems or where surface effects and long-range interactions distort equivalence, demanding ensemble-specific treatments.

While ensembles constitute an idealized abstraction, they distinctly scaffold the statistical mechanics landscape, stitching together microstates with broad thermodynamic landscapes. Their practical versatility and adaptability beget robust approaches for analyzing experimental data, forecasting system behaviors, and contextualizing divergent phenomena under a unified understanding of statistical thermodynamics.

This cornerstone framework perpetuates an expansive tapestry of applications, from crafting the laws of thermodynamics to pioneering contemporary pursuits encompassing quantum computing, nanoparticle synthesis, and complex biological networks—all rooted in the harmonious interplay of canonical, microcanonical, and grand canonical ensembles.

6.6 Connection between Statistical Mechanics and Thermodynamics

Statistical mechanics provides a powerful microscopic foundation for the macroscopic laws of thermodynamics, forming a bridge that links the collective behavior of individual atoms and molecules to observable large-scale properties. This connection is grounded in robust theoretical constructs that translate statistical averages into fundamental thermodynamic quantities, offering profound insights into system behaviors under varied conditions.

The essence of statistical mechanics lies in its ability to predict and explain thermodynamic properties by examining the distribution and dynamics of microstates. Each microstate represents a distinct configuration of a system's particles, characterized by specific positions and momenta. In thermodynamics, a system is described by macrostates, which are defined in terms of measurable quantities like temperature, volume, pressure, and entropy.

The bridge between the two is established through probabilistic methods: sta-

6.6. CONNECTION BETWEEN STATISTICAL MECHANICS AND THERMODYNAMICS

tistical mechanics utilizes probability distributions to determine the likelihood of a system being in any given microstate. By calculating expected values over these distributions, it derives macroscopic properties from averages of the microscopic details.

One of the foundational principles in this regard is the use of ensemble averages. An ensemble is a collection of virtual systems considered in all possible microstates consistent with a given macrostate. The ensemble averages of observables directly correspond to thermodynamic properties. For instance, the internal energy U in the canonical ensemble is given by:

$$U = \langle E \rangle = \sum_i P_i E_i$$

where P_i represents the probability of the system being in microstate i with energy E_i.

A core aspect of connecting statistical mechanics to thermodynamics is the statistical definition of entropy. In classical thermodynamics, entropy S is a measure of disorder or randomness. Statistically, entropy is defined by Boltzmann's relation:

$$S = k_B \ln \Omega$$

where Ω is the number of available microstates at a given energy level. This definition aligns directly with the second law of thermodynamics, which states that the entropy of an isolated system never decreases, signifying that systems evolve towards the most probable configuration with the maximum Ω.

This statistical underpinning unifies the macroscopic understanding of irreversible processes, underpinning concepts such as the arrow of time and the observation that natural processes progress towards thermodynamic equilibrium.

Thermodynamic properties typically present as averages in large systems, but fluctuations around these averages offer critical insights into system behavior, particularly in finite or small systems. Statistical mechanics quantifies these fluctuations through variance computation within ensembles, providing corrections to classical thermodynamics.

For example, in the canonical ensemble, the fluctuation in energy ΔE relates to the specific heat C_V by:

$$\langle(\Delta E)^2\rangle = k_B T^2 C_V$$

Such relations clarify the roles of microscopic variations in yielding macroscopic stability, inherent in thermodynamic limits. In tightly constrained systems or those experiencing phase transitions, fluctuations become pronounced, evidencing critical phenomena and cooperative effects inaccessible to classical descriptions alone.

Statistical mechanics further enriches the thermodynamic narrative by elucidating how systems exchange energy as heat and perform work at microscopic levels. It provides a rigorous basis for understanding free energy changes, connecting microscopic transitions to macroscopic constraints.

For instance, the Helmholtz free energy F is expressed in terms of partition functions within statistical mechanics:

$$F = -k_B T \ln Z$$

This relationship encapsulates the delicate balance between energy contributions and entropy, dictating spontaneous processes' direction and extent under constant temperature and volume, thus playing a pivotal role in chemical reactions and phase stability analysis.

Similarly, the Gibbs free energy G, adaptable to constant temperature and pressure conditions, ties directly to equilibrium and non-equilibrium phenomena, dictating reactions' feasibility and equilibrium positions in chemical systems.

The richness of statistical mechanics is vividly displayed in the study of phase transitions—the transformations between different states of matter. These phenomena exhibit nontrivial behavior at critical points, where traditional thermodynamic approaches falter.

By employing statistical mechanics, one can derive phase diagrams and understand critical exponents, universality classes, and scaling laws that characterize transitions. Renowned models, such as Ising and Potts models, demonstrate how simple microscopic interactions lead to complex macroscopic order phenomena including ferromagnetism and fluid criticality.

Phase transitions are intimately tied to free energy landscapes and are described through non-analyticities in generalized partition functions. Using techniques like mean-field theory and renormalization group analysis, statistical mechanics provides powerful tools to comprehend emergent phenomena

6.6. CONNECTION BETWEEN STATISTICAL MECHANICS AND THERMODYNAMICS

at and near critical points, fostering advances in fields ranging from condensed matter physics to biophysics.

The connection between statistical mechanics and thermodynamics extends gracefully into quantum realms, where indistinguishability and quantum correlations introduce novel thermodynamic properties unattainable via classical intuitions.

Quantum statistical mechanics, incorporating Fermi-Dirac and Bose-Einstein statistics, adapts classical thermodynamic laws to systems of indistinguishable particles. Fermi gases, for example, illustrate degeneracy pressure, offering insights into electronic structures of atoms, neutron stars, and quantum cells critical for elucidating high-energy density physics.

Quantum thermodynamics probes thermalization processes in closed quantum systems, explores quantum coherence at thermodynamic limits, and advances understanding in quantum information theory. Developments in this area challenge classical boundaries by exploiting quantum entanglement and coherence, shining light on engines and refrigerators operating beyond quasistatic limits.

Statistical mechanics enhances nonequilibrium thermodynamics by analyzing systems far from equilibrium—those dynamically adjusting to external perturbations. This has given rise to fluctuation theorems and Jarzynski equality, offering deep insights into work extraction and dissipation in small systems, vital for the burgeoning fields of nanoscale thermodynamics and biomolecular motors.

These advances illuminate how irreversible behaviors manifest at microscopic levels and relate intricate energy landscapes to macroscopic path-dependent phenomena, bridging classical thermodynamics with practical realities in biological systems and nanotechnology.

In laboratory and industrial settings, statistical mechanics underpins advances in material science, electronics, and energy technology. Predictive models of heat capacities, response functions, and transport coefficients derived from statistical distribution laws inform the design and optimization of materials and devices for renewable energy, information processing, and low-temperature physics.

Modern computational techniques such as molecular dynamics and Monte Carlo simulations, grounded in statistical mechanics, allow precise modeling of complex systems at atomic scales, yielding insights into catalysis, polymer science, and drug discovery, where classical approaches lack resolution.

The connection between statistical mechanics and thermodynamics reveals itself as a tapestry woven from probabilities at atomic scales to macroscopic laws that govern reality as we perceive it. Bridging this gap furnishes a comprehensive view of the natural world, empowering both theoretical exploration and practical innovation. Together, they encapsulate the harmony between the microscopic dance of particles and the macroscopic architectures of order and equilibrium, shaping our quantitative understanding of the physical universe and the myriad phenomena it unfolds.

6.7 Models and Approximations in Thermodynamics

Models and approximations in thermodynamics are essential tools for understanding complex systems. They facilitate the analysis of thermodynamic processes by simplifying real-world phenomena into manageable mathematical frameworks. These models allow scientists to predict the behavior of systems under various conditions and to develop insights that can guide experimental and industrial applications. This section delves into the commonly employed models and approximations, emphasizing ideal gas models, real gas models including Van der Waals and virial expansions, the use of lattice models, and mean-field theories for analyzing phase transitions.

Ideal Gas Model

The ideal gas model serves as the most fundamental representation of gas behavior. It is based on a series of simplifying assumptions: gas particles are considered point-like with no intermolecular forces, they move in straight lines until colliding with each other or the walls, and all collisions are perfectly elastic. The ideal gas law is expressed as:

$$PV = nRT$$

where P is the pressure, V the volume, n the number of moles, R the ideal gas constant, and T the temperature in kelvins.

Despite its simplicity, the ideal gas law provides a remarkably accurate description of gases under a wide range of conditions, especially at high temperatures and low pressures where intermolecular forces and molecular volumes

6.7. MODELS AND APPROXIMATIONS IN THERMODYNAMICS

become negligible. This model finds applications in diverse fields ranging from meteorology to engineering, underpinning technologies like air conditioning systems and internal combustion engines.

Real Gas Models

In practical applications, deviations from ideal gas behavior become noticeable at high pressures and low temperatures. To address these deviations, real gas models incorporate interactions between molecules and their finite volumes. The Van der Waals equation represents one of the earliest attempts to correct the ideal gas law:

$$\left(P + \frac{n^2 a}{V^2}\right)(V - nb) = nRT$$

Here, a and b are constants specific to each gas, accounting for the attractive forces and molecular size, respectively. The Van der Waals model predicts phenomena such as the critical point and liquid-vapor equilibrium, offering valuable insights into phase transitions and condensation processes.

Beyond Van der Waals, the virial expansion provides a systematic approach to address non-ideal gases through a series of corrections to the ideal gas law:

$$\frac{PV}{nRT} = 1 + \frac{B(T)}{V} + \frac{C(T)}{V^2} + \cdots$$

where $B(T)$, $C(T)$, etc., are the virial coefficients, functions of temperature derived from intermolecular potential models. These coefficients allow for the representation of interactions more accurately over a range of conditions, enhancing calculations of compressibility factors and aiding in the design of processes such as gas liquefaction.

Lattice Models in Statistical Mechanics

Lattice models, particularly the Ising and Heisenberg models, are instrumental in studying magnetic systems and phase transitions. In these simplifications, a lattice represents regular grid-like structures wherein nodes denote possible states, such as spins in magnetic materials. The Ising model considers spins that can adopt up or down states, illustrating ferromagnetism through interactions between neighboring spins:

$$H_{\text{Ising}} = -J \sum_{\langle i,j \rangle} s_i s_j - h \sum_i s_i$$

where J is the exchange interaction constant, s_i and s_j are spin variables, and h is an external magnetic field. Its simplicity makes it ideal for demonstrating spontaneous magnetization below a critical temperature, providing insights into order-disorder transitions.

The Heisenberg model extends this approach by incorporating three-dimensional vector spins, elaborating on quantum mechanical interactions. These models find relevance in materials science and condensed matter physics, serving as foundations for understanding cooperative phenomena and criticality.

Mean-Field Theory and Phase Transitions

Mean-field theory offers a significant approximation method, simplifying the complex many-body interactions in a system by averaging effects over the field. This technique is applied widely to estimate critical exponents, transition phenomena, and to predict macroscopic behavior from microscopic theories.

In the Landau theory of phase transitions, mean-field approximations consider the free energy expansion in terms of an order parameter ϕ:

$$F(\phi) = F_0 + a\phi^2 + b\phi^4 + \cdots$$

Here, coefficients such as a change sign at critical temperatures, indicating transitions, while higher-order terms stabilize the process. This framework is adept at exploring ferroelectricity, superconductivity, and liquid crystals, reflecting how symmetries and dimensionality influence phase changes.

Heat Capacity Models for Solids

Thermodynamic modeling extends into solid-state physics, where Debye and Einstein models predict heat capacities. The Einstein model treats atoms as independent harmonic oscillators, while the Debye model accounts for collective modes of lattice vibrations or phonons, manifesting as:

$$C_V = 9nR \left(\frac{T}{\Theta_D} \right)^3 \int_0^{\frac{\Theta_D}{T}} \frac{x^4 e^x}{(e^x - 1)^2} \, dx$$

6.7. MODELS AND APPROXIMATIONS IN THERMODYNAMICS

where Θ_D is the Debye temperature. These models successfully describe low-temperature behaviors where quantum effects dominate, aiding in material exploration and the development of thermoelectric applications.

Approximate Models in Chemical Thermodynamics

In chemical thermodynamics, models and approximations are vital for evaluating reaction thermodynamics, phase equilibria, and solution behaviors. Raoult's and Henry's laws describe ideal and dilute solutions respectively, guiding predictions of vapor pressure and solubility.

For electrolyte solutions, Debye-Hückel theory offers approximations for ionic interactions, illuminating electrochemical processes and biological systems' functionality.

Kinetic Theory and Transport Phenomena

The kinetic theory of gases, another central model, describes gas molecules' motion, linking macroscopic properties like pressure and viscosity to molecular dynamics. For instance, the Chapman-Enskog theory provides approximate solutions to the Boltzmann equation, elucidating transport phenomena including diffusion and thermal conductivity in gases.

Through kinetic models, it is possible to derive Navier-Stokes equations in fluid dynamics, extending thermodynamic principles into dynamical realms. These approaches underpin fluid mechanics, meteorology, and aerospace engineering, offering robust frameworks for analyzing laminar and turbulent flows.

Limitations and Scope of Thermodynamic Models

While models and approximations facilitate invaluable insights, they inherently carry limitations. Many rely on simplifying assumptions such as non-interacting particles or uniform fields, which may overlook complex interactions and non-equilibrium behaviors.

The ongoing development of computational methods, such as molecular dynamics and Monte Carlo simulations, increasingly supplements classical models, allowing exploration of systems with greater resolution and fewer constraints. However, analytical models sustain their significance by providing fundamental insights, guiding initial explorations, and yielding qualitative predictions that continue to illuminate the vast landscape of thermodynamic phenomena.

Models and approximations embody the heart of thermodynamics, transforming the intricacies of microscopic interactions into comprehensible macroscopic descriptions. They empower predictions, drive innovations, and integrate theoretical advances with experimental data, continuing to shape our understanding and application of thermodynamic principles across diverse scientific and engineering landscapes.

Chapter 7

Spectroscopy and Photochemistry

Spectroscopy and photochemistry delve into the interaction of light with matter, offering profound insights into molecular structure and dynamical processes. This chapter explores various spectroscopic techniques, such as infrared, UV-Vis, and NMR, elucidating how they probe vibrational, electronic, and magnetic properties to characterize substances. The principles governing light absorption, emission, and scattering reveal the quantized nature of energy levels within molecules. In photochemistry, the focus shifts to the chemical changes induced by light, highlighting the mechanisms of photoexcitation and subsequent photophysical events. Together, these disciplines furnish a robust toolkit for deciphering both static and dynamic aspects of molecular identity and reactivity.

7.1 Principles of Spectroscopy

Spectroscopy is an analytical technique employed to study the interaction between electromagnetic radiation and matter. This interaction allows for the identification and characterization of substances at the molecular and atomic levels. The foundation of spectroscopy is built upon two critical phenomena: the absorption and emission of electromagnetic radiation. These interactions facilitate the understanding of molecular structures, chemical compositions, and dynamic processes occurring within substances.

The principles governing spectroscopy are rooted in the behavior of electromagnetic radiation, which travels in wave-like patterns characterized by distinct wavelengths and frequencies. The electromagnetic spectrum encompasses all wavelengths of electromagnetic radiation, ranging from radio waves with long wavelengths and low frequencies, to gamma rays with short wavelengths and high frequencies. Within this spectrum, visible light occupies a narrow region, delineating wavelengths detectable by the human eye, approximately 400 to 700 nm.

Spectroscopy exploits the quantized nature of energy levels within atoms and molecules. When electromagnetic radiation impinges upon matter, radiation energy can be absorbed if its energy matches the difference between quantized energy levels in a molecule or atom. This process of energy absorption elevates the system to an excited state. Upon returning to a ground state, energy can be re-emitted as electromagnetic radiation, manifesting as emission spectra.

A diagrammatic representation of the electromagnetic spectrum is instrumental in visualizing the various regions and corresponding processes:

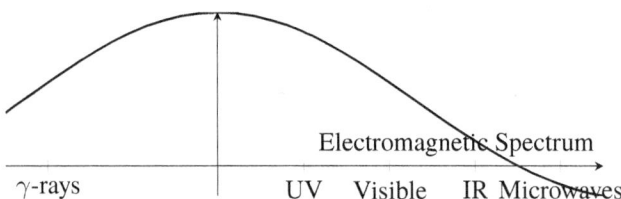

The quantization of energy levels is explained by quantum mechanics, which provides profound insights into how molecular and atomic transitions occur. Quantum theory posits that atoms and molecules have discrete energy levels. The energy difference between these levels determines the specific wavelengths of radiation absorbed or emitted. This quantized behavior is a direct consequence of the wave-particle duality of light and matter—principles articulated in foundational quantum theories.

The interaction of light with matter can result in various types of spectroscopic processes. When photons supply energy sufficient to induce electronic transitions in atoms, we encounter electronic spectroscopy. Molecules absorb ultra-

7.1. PRINCIPLES OF SPECTROSCOPY

violet or visible light, promoting electrons to higher energy orbitals. Vibration and rotational transitions involve lower energy changes, typically observed in the infrared region—a principle exploited in infrared spectroscopy.

To illustrate the concept of energy transitions, consider the following:

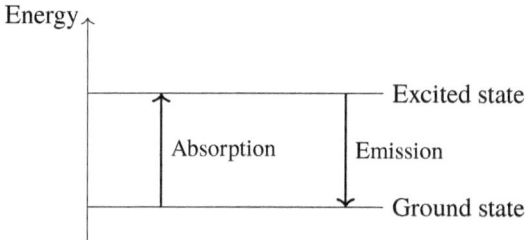

Within the electromagnetic spectrum, different regions hold suitability for specific types of spectroscopy. X-rays, with their high energy and short wavelengths, are powerful for elucidating atomic structures in crystallography; infrared radiation is adept at detecting vibrational transitions in molecules, providing insights into bond strengths and chemical environments; and visible light engenders color perception and electronic spectroscopic analysis.

The absorption of light by simple molecules results in a spectrum unique to each species. As each molecule has its characteristic energy level transitions, the specificity allows us to identify substances through their spectral fingerprints. The law governing this phenomenon is encapsulated in Beer-Lambert's law, which correlates absorbance to concentration, path length, and molar absorptivity.

$$A = \epsilon c l$$

Where A represents absorbance, ϵ denotes the molar absorptivity, c is the concentration, and l signifies the path length through the sample. This linear relationship is foundational in quantitative spectroscopic analysis, enabling concentration determinations by measuring absorbance.

Taking a closer look at absorption and emission spectroscopy, these techniques, although rooted in the interaction of light with matter, are delineated by their differing processes. Absorption spectroscopy measures transitions from lower to higher energy states as light passes through a sample, while emission spectroscopy studies radiation emitted by a sample when it descends from higher to lower energy states.

A practical implication of these spectroscopic techniques lies in their wide applicability across scientific domains. In chemistry, spectroscopy elucidates structural details of organic compounds, detects impurities, and assists in reaction monitoring. In astrophysics, identifying the composition of distant stars and galaxies is achievable through spectroscopy by analyzing the electromagnetic radiation emitted.

Additionally, advancements in technology have enabled the refinement of spectroscopic instruments enabling high precision and resolution measurements. Fourier-transform spectroscopy leverages mathematical transformations to convert raw data into interpretable spectral data, enhancing signal processing capabilities.

In summary, the principles of spectroscopy offer profound insights into the molecular and atomic world. The methodical analysis of light-matter interactions yields invaluable data, proving integral in myriad applications from fundamental research to industrial analytics. Understanding these principles enhances our capacity to explore the universe in dimensions previously obscured to direct observation.

7.2 Infrared and Raman Spectroscopy

Infrared (IR) and Raman spectroscopy are complementary techniques employed to examine vibrational transitions in molecules. These spectroscopic methods provide crucial data for understanding molecular structures, identifying functional groups, and exploring the dynamic behavior of chemical species. Both techniques hinge upon the interaction between electromagnetic radiation and molecular vibrations, yet they differ in their fundamental principles and the nature of the interactions.

Infrared Spectroscopy

Infrared spectroscopy involves the absorption of infrared radiation by molecules, which causes changes in their vibrational energy levels. When IR radiation is absorbed, certain vibrational modes within the molecule are excited, leading to an increase in vibrational energy. The vibrational modes that are active in IR spectroscopy arise from changes in the dipole moment of the molecule; hence, polar molecules often exhibit strong IR absorption bands.

The IR spectrum typically spans the mid-infrared region (4000–400 cm^{-1}), where vibrations involving stretching and bending of chemical bonds are observed. Two main types of vibrational transitions are associated with IR absorption: stretching (which involves changes in bond length) and bending (which involves changes in bond angles).

A flowchart of the fundamental process in infrared spectroscopy is beneficial for clarifying this concept:

Molecular vibrations can be further categorized into symmetric and asymmetric stretching, as well as scissoring, rocking, wagging, and twisting—each characterized by distinctive energy changes. The position and intensity of these absorptions can be used to identify the presence of specific functional groups within a molecule. For example, the carbonyl group (C=O) exhibits a strong absorption band typically around 1700 cm^{-1}.

The instrument used to record an IR spectrum is an IR spectrophotometer, which measures the amount of IR radiation absorbed at various wavelengths. The result is a plot of transmittance or absorbance versus wavenumber (reciprocal of the wavelength), creating a unique spectral fingerprint for each compound.

Raman Spectroscopy

Contrary to IR spectroscopy, Raman spectroscopy is based on the scattering of light. It involves the inelastic scattering of photons, known as the Raman effect, discovered by C. V. Raman in 1928. When monochromatic light—usually from a laser—is incident on a molecule, most of it is elastically scattered (Rayleigh scattering); however, a small portion is inelastically scattered, resulting in frequency shifts corresponding to vibrational energy levels of the molecules.

The key to Raman spectroscopy is the polarizability change of the molecule rather than changes in dipole moment. Non-polar molecules can be quite active in Raman spectra, even if they are IR inactive, making Raman spectroscopy a complementary technique to IR spectroscopy.

The process of Raman scattering can be elucidated as follows:

CHAPTER 7. SPECTROSCOPY AND PHOTOCHEMISTRY

The Raman spectrum is derived from Stokes shifts, where scattered photons have less energy than the incident light, and Anti-Stokes shifts, which occur when scattered photons have more energy (generally weaker than Stokes lines).

Comparative Analysis and Applications

Infrared and Raman spectroscopy are powerful techniques for molecular analysis. While IR spectroscopy is superior at identifying polar functional groups due to their prominent absorption bands, Raman spectroscopy excels in analyzing symmetrical, non-polar molecules that show intense Raman signals.

In practical applications, these techniques are indispensable in various fields:

- **Chemistry and Biotechnology:** IR and Raman provide insights into molecular compositions and classifications of compounds based on functional groups. They are instrumental in substance identification, determining stereochemical arrangements, and analyzing biochemical samples, such as proteins and nucleic acids.

- **Material Science:** Detection and quantification of molecular defects in solids, assessment of crystallinity, and analysis of stress and strain in materials are facilitated by these spectroscopic techniques. Raman, in particular, is well-suited for studying carbonaceous materials, such as graphene, due to its sensitivity to carbon-carbon bonds.

- **Pharmaceuticals:** Monitoring the stability of drugs, analyzing formulations and polymorphs, and performing quality control on pharmaceutical products are accomplished using IR and Raman spectroscopy. These techniques aid in the fast, non-destructive testing of pharmaceutical substances.

Complementary Nature and Challenges

The complementary nature of IR and Raman spectroscopy manifests in the distinct selection rules governing each technique. Consequently, a comprehensive understanding of molecular vibration often requires combining both methods. The simultaneous analysis enriches data interpretation and ensures a more detailed molecular characterization.

Despite their prowess, certain challenges obstruct their use. In IR spectroscopy, sample preparation often necessitates thin films or pressed pellet forms, while Raman spectroscopy suffers from fluorescence interference and requires carefully selected laser sources to minimize sample damage.

Advancements in spectroscopic instrumentation, including Fourier-transform IR (FTIR) and confocal Raman microscopy, continue to mitigate these drawbacks. These technologies have enhanced sensitivity, resolution, and the ability to examine diverse sample matrices with minimal preparation, extending the applicability of IR and Raman spectroscopy across various scientific disciplines.

Infrared and Raman spectroscopy synergistically enhance our ability to explore the vibrational landscape of molecules. By leveraging the detailed molecular information provided through each unique spectral profile, researchers and scientists are empowered to solve complex problems within both academic and industrial settings. These spectroscopic tools persist as fundamental components of the analytical arsenal, advancing scientific inquiry and practical applications across diverse fields.

7.3 Ultraviolet-Visible (UV-Vis) Spectroscopy

Ultraviolet-Visible (UV-Vis) spectroscopy is a widely utilized analytical technique that exploits the absorption of ultraviolet and visible light by molecules. It is particularly effective in exploring electronic transitions, providing valuable insights into the concentration, composition, and behavior of chemical species in solution. This section delves into the principles, instrumentation, applications, and interpretative methods of UV-Vis spectroscopy, emphasizing its relevance in both research and applied sciences.

Principles of UV-Vis Spectroscopy

UV-Vis spectroscopy operates on the principle of electronic transitions within molecules. When molecules absorb ultraviolet or visible light, electrons are excited from the ground state to higher energy electronic states. The energy absorbed corresponds to the difference between the ground and excited states, which falls within the ultraviolet (200-400 nm) and the visible (400-700 nm) regions of the electromagnetic spectrum.

Electromagnetic radiation can be described in terms of its wavelength (λ), frequency (ν), and energy (E), related by the equations:

$$E = h\nu = \frac{hc}{\lambda}$$

where h is Planck's constant and c is the speed of light.

Electronic excitation typically involves transitions from bonding (σ), non-bonding (n), or π orbitals to anti-bonding (σ^* or π^*) orbitals. These transitions include:

- $\sigma \to \sigma^*$: Involve high-energy absorptions, usually occurring in the vacuum-UV region and less relevant in UV-Vis spectroscopy.

- $n \to \sigma^*$: Seen in molecules with lone pairs, such as oxygen and nitrogen-containing compounds.

- $\pi \to \pi^*$: Characteristic of unsaturated systems such as alkenes, dienes, and aromatic compounds.

The characteristic absorption bands provide information about the electronic structure of molecules and contribute to determining their identity and concentration.

Instrumentation

The basic components of a UV-Vis spectrophotometer include a radiation source, a monochromator for wavelength selection, a sample holder, and a detector. Advanced models incorporate digital processing capabilities for data acquisition and analysis.

- **Radiation Source**: Typically, deuterium lamps are used for the UV region, while tungsten lamps cater to the visible spectrum. The lamps emit continuous spectra, offering a range of wavelengths for examination.

- **Monochromator**: This device isolates specific wavelengths from the emitted light, often achieved using prisms or diffraction gratings, enabling the precise measurement of absorbance at individual wavelengths.

- **Sample Holder**: Transparent cells or cuvettes, usually made of quartz or glass, hold the sample. The path length of the cuvette is a critical parameter, factored into calculations governing absorbance.

- **Detector**: Photomultiplier tubes or photodiodes convert the transmitted light into an electrical signal proportional to the light intensity, facilitating quantification.

- **Display/Output**: Modern instruments feature computerized displays that graph absorbance or transmittance as a function of wavelength, generating spectra for analysis.

Beer-Lambert Law

The quantitative determination of a sample's concentration in UV-Vis spectroscopy relies on the Beer-Lambert Law, which states:

$$A = \epsilon \cdot c \cdot l$$

where A is absorbance, ϵ is the molar absorptivity (L mol^{-1} cm^{-1}), c is concentration (mol L^{-1}), and l is path length (cm). The linear relationship between absorbance and concentration allows for calibration and quantitation using standard solutions.

Applications of UV-Vis Spectroscopy

UV-Vis spectroscopy's versatility endears it to numerous scientific disciplines, supported by its rapid, reliable, and non-destructive analysis capabilities. Key applications include:

- **Chemical Analysis**: Determining the concentration of molecules with chromophores, identifying unknown compounds, and assessing the purity of chemical substances are common uses in analytical chemistry.

- **Biochemical Studies**: Monitoring nucleic acids and proteins is paramount in biochemistry. For instance, nucleic acids absorb strongly at 260 nm due to the $\pi \rightarrow \pi^*$ transitions of their aromatic bases, whereas proteins absorb at 280 nm, a result of the presence of aromatic amino acids like tryptophan and tyrosine.

- **Industrial Applications**: UV-Vis spectroscopy supports quality control in manufacturing processes, from pharmaceuticals, where it measures active ingredient concentration, to food and beverage industries, where it assesses the purity and concentration of colorants and additives.

- **Environmental Monitoring**: Water quality assessment frequently employs UV-Vis spectroscopy to measure pollutants, such as nitrates and phosphates, by detecting characteristic absorption peaks.

Example Interpretation

To elucidate how UV-Vis spectra can reveal molecular details, consider benzene, a simple aromatic compound. Its UV-Vis spectrum shows a prominent peak near 256 nm, attributable to $\pi \rightarrow \pi^*$ electronic transitions in its conjugated ring system. Additionally, smaller shoulders around 200 nm suggest higher energy transitions.

Interpreting spectra involves analyzing peak positions (wavelengths), intensities (absorbances), and shapes (broad or sharp). Comparisons against reference spectra or databases aid identification. Furthermore, spectrum deconvolution can resolve overlapping peaks, enabling detailed molecular analysis.

Challenges and Advancements

While UV-Vis spectroscopy is robust, certain challenges must be addressed. The method is inherently reliant on clear, colorless solvents that do not absorb strongly in the UV-Vis range, limiting solvent choice. Additionally, the presence of interfering substances can skew results, necessitating meticulous sample preparation and instrument calibration.

Recent advancements have countered these issues through innovations like:

- **Diffuse Reflectance and Attenuated Total Reflectance (ATR)**: These techniques facilitate the analysis of solid and turbid samples, expanding the applicability of UV-Vis spectroscopy to new domains.

- **Miniaturized and Portable Devices**: Enhanced portability and field-deployment capabilities meet the demand for on-site analysis. These devices, often coupled with smartphone technology, provide real-time data and rapid decision support.

- **Coupling with Chromatography**: The integration of UV-Vis detectors with high-performance liquid chromatography (HPLC) enhances separation and detection, providing synergy in complex mixture analysis, common in pharmacokinetics and metabolomics studies.

UV-Vis spectroscopy remains a foundational technique in analytical science. By delivering insights into electronic structures and providing a mechanism for concentration assessment, it is an indispensable tool in both academic and industrial research. As technology evolves, the scope and precision of UV-Vis analysis continually advance, further solidifying its role as an integral component of the analytical toolkit.

7.4 Nuclear Magnetic Resonance (NMR) Spectroscopy

Nuclear Magnetic Resonance (NMR) Spectroscopy is a powerful analytical technique pivotal in elucidating the structure, dynamics, and interactions of molecules. It exploits the magnetic properties of certain nuclei, offering unparalleled insights into molecular architecture and the environment surrounding these nuclei. Its application spans diverse fields, including chemistry, biochemistry, medicine, and materials science, owing to its ability to provide detailed information about molecular structure in both solid and liquid states.

Principles of NMR Spectroscopy

NMR spectroscopy is based on the principle of nuclear spin. Certain atomic nuclei possess an intrinsic spin and associated magnetic moment. When placed in a strong external magnetic field, these nuclei align with or against the field, corresponding to different energy states. Nuclei with a non-zero spin angular momentum ($I \neq 0$), such as ^{1}H, ^{1}3C, ^{1}5N, and ^{3}1P, are NMR-active and most commonly studied.

The fundamental equation governing the energy difference (ΔE) between nuclear spin states is given by:

$$\Delta E = \hbar \gamma B_0$$

where $\hbar$ is the reduced Planck's constant, γ is the gyromagnetic ratio (a constant unique to each type of nucleus), and B_0 is the external magnetic field strength.

When the appropriate radiofrequency (RF) radiation resonates with this energy gap, nuclei are excited from a lower to higher spin state. As they relax back, they emit RF signals that are captured and processed to yield NMR spectra.

A flowchart representing the NMR process is critical for understanding the sequence of events leading to spectra generation:

NMR Spectroscopy Process — Sample in External Magnetic Field — Detection of Emitted RF Signal — Fourier Transform — NMR Spectrum Generation / Application of RF Pulse — Nuclear Excitation

Chemical Shift: One of the most crucial parameters in NMR spectroscopy is the chemical shift, denoted as δ, which quantifies the resonance frequency of a nucleus relative to a reference standard, typically TMS (tetramethylsilane). Chemical shifts are reported in parts per million (ppm) and reflect the electronic environment around each type of nucleus, influenced by factors such as electronegativity, hybridization, and magnetic anisotropy within a molecule.

$$\delta = \left(\frac{\nu - \nu_{\text{ref}}}{\nu_{\text{ref}}} \right) \times 10^6$$

Spin-Spin Coupling: NMR spectra are characterized by splitting patterns resulting from spin-spin coupling between adjacent nuclei. The coupling constant, J, measured in Hertz (Hz), indicates the extent of interaction and helps deduce the number of adjacent protons, providing insight into molecular topology.

Relaxation: The return of nuclei to equilibrium following perturbation, encompassing T_1 (longitudinal) and T_2 (transverse) relaxation processes, further characterizes the NMR signal and informs on molecular dynamics.

Instrumentation and Techniques

NMR spectrometers comprise strong superconducting magnets, RF transmission and detection systems, and sophisticated digital processing units. Key techniques include:

- **Proton NMR (^{1}H NMR):** The most common form, providing information on hydrogen-containing compounds. Chemical shifts typically range from 0 to 12 ppm, with characteristic regions for different functional groups.

- **Carbon-13 NMR (^{13}C NMR):** Offers details on the carbon skeleton, typically requiring signal enhancement techniques like broadband decoupling or Nuclear Overhauser Effect (NOE) to address lower sensitivity due to the ^{13}C natural abundance.

- **Multidimensional NMR:** Advances, such as 2D NMR (COSY, NOESY, HSQC), enable complex molecules study by correlating

7.4. NUCLEAR MAGNETIC RESONANCE (NMR) SPECTROSCOPY

interactions between nuclei, revealing detailed spatial arrangements.

- **Solid-State NMR:** Utilizes techniques like magic angle spinning (MAS) to mitigate anisotropic interactions in solids, enhancing resolution.

Applications of NMR Spectroscopy

NMR spectroscopy's applications are extensive and continually expanding, enriching numerous scientific disciplines:

- **Structural Elucidation:** In organic and inorganic chemistry, NMR determines structural details, including stereochemistry, conformation, and tautomerism. It identifies isomers, reaction intermediates, and products, providing vital checks on synthetic processes.

- **Biochemistry and Structural Biology:** NMR uncovers biomolecular structures at atomic resolution. Protein-ligand interactions, folding, and dynamics are explored through tailored techniques like isotope labeling, pivotal in drug design and enzyme mechanism studies.

- **Materials Science:** Analysis of polymers, nanomaterials, and complex metallic systems is achieved by employing solid-state NMR techniques, offering information on phase composition, crystallinity, and defect sites.

- **Medical Imaging:** NMR is the underlying science of Magnetic Resonance Imaging (MRI), revolutionizing medical diagnostics by providing non-invasive, high-contrast images of soft tissues.

Example Interpretation

Consider a simple molecule, ethanol, CH_3CH_2OH. The ^1H NMR spectrum of ethanol exhibits characteristic signals:

- A triplet at about 1 ppm corresponding to the CH_3 group.

- A quartet around 3.6 ppm from the CH_2 group, coupled to the adjacent methyl hydrogens.

- A broad singlet arising from the OH proton, often appearing variable in position due to hydrogen bonding.

Through analysis of chemical shifts, multiplicity, and integration, the molecular structure and functional groups are inferred.

Challenges and Innovations

While NMR offers detailed molecular insights, challenges such as sensitivity and resolution remain, impeding its applicability to low-concentration samples or large biomolecules. Innovations addressing these barriers include:

- **Hyperpolarization Techniques:** Methods like dynamic nuclear polarization (DNP) substantially enhance signal strength, expanding the potential to study complex biosystems and real-time processes.

- **Cryoprobes:** Improved detection sensitivity by cooling the RF coil and reducing thermal noise, proving beneficial for high-resolution studies.

- **In-Situ NMR:** Allows analysis of chemical reactions as they occur, providing insights into kinetics and mechanism directly in the reaction environment.

Ultimately, NMR spectroscopy is an indispensable analytical tool offering unparalleled insights across scientific domains. By providing detailed molecular information in both static and dynamic contexts, NMR enables the resolution of complex problems, fueled by ongoing technological advancements. Its versatility and depth render it a cornerstone of both research and application, continually ushering in new scientific understanding.

7.5 Mass Spectrometry

Mass spectrometry (MS) is an analytical technique of paramount importance, utilized for determining the mass-to-charge ratio (m/z) of ions. It provides profound insights into the composition of chemical substances, facilitating their qualitative and quantitative analysis. Mass spectrometry is indispensable across various scientific disciplines, offering unparalleled capabilities in analyzing complex mixtures, identifying unknown compounds, and elucidating structural and isotopic details.

Principles of Mass Spectrometry

The core of mass spectrometry lies in the conversion of molecules into ions, which are then separated based on their mass-to-charge ratios. These ions produce signals that are recorded and interpreted to yield a mass spectrum—a distinct fingerprint used for molecular identification and quantification.

The fundamental processes of mass spectrometry are depicted in the following stages:

- **Ionization**: The initial step involves converting neutral molecules into ions. Depending on the technique used, ionization can be hard (e.g., electron ionization, causing fragmentation) or soft (e.g., electrospray ionization, preserving molecular integrity).

- **Mass Analysis**: Ions generated are directed into a mass analyzer, where they are sorted based on their mass-to-charge ratios. Different types of mass analyzers include quadrupole, time-of-flight (TOF), and orbitrap, each with unique resolutions and sensitivities.

- **Detection**: After separation, ions reach the detector. The time or current associated with ion arrival is converted into mass-to-charge readings, which are translated into a spectrum.

A flowchart illustrating the mass spectrometry process elucidates this sequence:

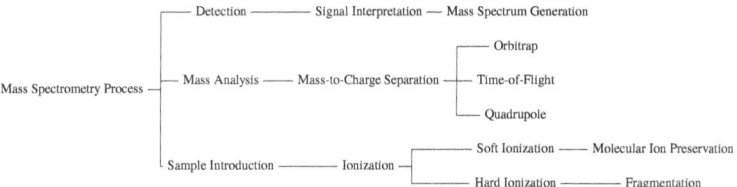

Types of Ionization Techniques

- **Electron Ionization (EI)**: Common in gas chromatography-mass spectrometry (GC-MS), EI involves bombarding gaseous molecules with high-energy electrons, leading to extensive fragmentation. The resulting spectra are rich in fragment ions, aiding structural elucidation.

- **Electrospray Ionization (ESI)**: Widely used in liquid chromatography-mass spectrometry (LC-MS), ESI gently ionizes large biomolecules, producing multi-charged ions by creating a fine spray of charged droplets in an electric field.

- **Matrix-Assisted Laser Desorption/Ionization (MALDI)**: Useful for analyzing large biomolecules like proteins and polymers, MALDI uses a laser to ionize samples embedded in a crystalline matrix, minimizing fragmentation.

Mass Analyzers and Their Applications

- **Quadrupole Mass Analyzer**: Utilizes oscillating electric fields to filter ions by mass. It offers rapid scanning and robustness, making it ideal for routine analyses and quantification in MS.

- **Time-of-Flight (TOF)**: Measures the time taken by ions to travel a fixed distance after acceleration by an electric field. TOF analyzers provide high resolution and mass range, proving essential for complex mixture analysis, particularly when coupled with MALDI.

- **Orbitrap**: Known for its high resolution and accuracy, the orbitrap analyzer traps ions in an electrostatic field, and mass is determined from frequency translations. It is predominantly used in studies requiring precise mass measurements and complex composition analysis.

Interpretation of Mass Spectra

Interpreting mass spectra involves analyzing the peaks corresponding to different ions. The most intense peak is designated as the base peak, commonly assigned an intensity of 100%. The molecular ion peak (M^+) represents the intact ion of the molecule, essential for determining molecular weight.

In the spectrum, isotopic patterns and fragment peaks provide further insight. For example, Cl and Br isotopes contribute characteristic patterns due to their natural abundance differences, aiding in the determination of halogenated compounds.

The fragment ion peaks, arising from the cleavage of covalent bonds, help elucidate structural components. Common fragmentations, such as the McLafferty rearrangement and alpha-cleavage, are crucial for deducing chemical structures.

Applications of Mass Spectrometry

Mass spectrometry's versatility translates to a myriad of scientific applications:

- **Proteomics and Metabolomics**: MS allows for the high-throughput analysis of proteins and metabolites, facilitating the identification and quantification of biological molecules. Techniques like tandem MS (MS/MS) and isotopic labeling further enhance its capability to analyze complex biological matrices.

- **Pharmaceutical Analysis**: Drug development and quality control deploy MS for purity testing, degradation studies, and bioavailability assessments. It is instrumental in pharmacokinetics and drug metabolism studies, allowing precise tracking of active ingredients and metabolites.

- **Environmental and Forensic Applications**: MS identifies pollutants in environmental samples, such as persistent organic pollutants (POPs), and aids forensic investigations in detecting trace evidence and illicit substances.

- **Material Science**: Elemental and isotopic analyses in materials science draw on MS to study metallic alloys, semiconductors, and nanomaterials, aiding material characterization and advancement.

Challenges and Innovations

While mass spectrometry is incredibly informative, challenges such as ion suppression in complex matrices, the need for skilled interpretation, and instrument costs are notable. To address these, technological advancements focus on improving throughput, resolution, and sensitivity, including:

- **Coupling with Chromatography**: Techniques like GC-MS and LC-MS effectively separate complex mixtures, leading to more precise and informative analyses.

- **Enhanced Ionization Techniques**: The development of ionization methods such as desorption electrospray ionization (DESI) and paper spray, allowing ambient and open-air ionization, broadens MS applicability and operational flexibility.

- **Data Analysis Software**: Advanced algorithms and software improve the deconvolution and interpretation of MS data, enabling automated identification and quantification, crucial for large-scale studies in complex fields like proteomics.

Mass spectrometry stands as a cornerstone analytical method capable of resolving intricate questions across a spectrum of scientific disciplines. Its ability to generate detailed molecular profiles makes it an indispensable tool in both research and applied science environments. As instrumentation and methodologies continue to advance, MS is poised to unlock even deeper understanding in the fields it underpins, continually broadening the horizons of analytical capabilities.

7.6 Fluorescence and Phosphorescence

Fluorescence and phosphorescence are photoluminescent phenomena that occur when a substance absorbs light energy and subsequently re-emits it. The processes are distinguished primarily by the timescales of emission and the electronic states involved. Both mechanisms are critical in fields ranging from analytical chemistry and materials science to biological imaging and environmental sensing, offering insights into molecular dynamics and environments.

Fundamental Principles

Fluorescence and phosphorescence are types of luminescence, whereby molecules emit light following excitation by electromagnetic radiation. When molecules absorb photons, they are excited from a ground electronic state (S_0) to an excited electronic state (S_1 or S_2).

- **Fluorescence**: Represents a rapid emission process, typically occurring within nanoseconds, as electrons return to the ground state from an excited singlet state ($S_1 \rightarrow S_0$). Fluorescent emissions cease almost immediately after the excitation source is removed.

- **Phosphorescence**: Involves transitions from an excited triplet state (T_1) to a ground singlet state (S_0), which are 'forbidden' and therefore slower, causing emissions to persist, often from milliseconds to hours.

The efficiency of fluorescence is quantified by the fluorescence quantum yield, a ratio of emitted to absorbed photons, reflecting the effectiveness of the process.

Factors Influencing Fluorescence

- **Conjugation and Aromaticity**: Molecules with extended conjugation systems or aromatic groups often exhibit strong fluorescence due to stable $\pi \to \pi^*$ transitions, as seen in compounds like fluorescein and rhodamine.

- **Solvent Effects**: Solvents can influence fluorescence by affecting molecular stability and electronic distribution. Polar solvents, through solvent polarity, can stabilize excited states differently than non-polar solvents.

- **Temperature and pH**: These alter molecular interactions and can quench fluorescence by facilitating non-radiative decay pathways.

- **Quenching Agents**: Presence of halogens or paramagnetic species can result in fluorescence quenching either collisional or static, which reduces emission by providing non-radiative pathways.

Applications of Fluorescence

Fluorescence's sensitivity and specificity render it a tool of choice in numerous applications:

- **Biological Imaging**: Fluorescent dyes and proteins (e.g., GFP) aid visualization of cellular structures and dynamics, facilitating studies in cell biology and medical diagnostics.

- **Chemical Sensors**: Fluorescent sensors detect and quantify ions, molecules, or changes in the environment (pH, temperature) with high selectivity.

- **Environmental Monitoring**: Fluorescence detects pollutants or biomolecules in water and soil, offering sensitive and rapid environmental assessments.

- **Molecular Dynamics**: Time-resolved fluorescence spectroscopy elucidates protein folding, molecular conformations, and interactions within nanoseconds.

Phosphorescence and Its Applications

Phosphorescence, with delayed emission, finds niche applications where prolonged afterglow is advantageous:

- **Security Markers**: Used in anti-counterfeit features, phosphorescent materials provide security markings visible under specific conditions.

- **Glow-in-the-Dark Products**: Items such as emergency signage and novelty items exploit phosphorescent materials like strontium aluminate for visible emission without continued excitation.

- **Biological Probes**: Phosphorescent probes measure oxygen levels in biological systems, capitalizing on emission lifetime changes in different oxygen environments.

Quenching and Environmental Effects

Both fluorescence and phosphorescence are susceptible to quenching, a process where emission is diminished due to non-radiative pathways. Quenching can arise from external factors such as collisional interactions or internal factors like energy transfer between system components. Examining quenching effects provides insights into molecular interactions and proximate environments.

Innovations and Future Directions

Innovative techniques have expanded the applications of fluorescence and phosphorescence beyond traditional boundaries:

- **Super-Resolution Microscopy**: Techniques like STED and PALM leverage photoluminescence for imaging beyond the diffraction limit, unraveling finer cellular details.

- **Fluorescent Lifetime Imaging**: Fluorescence lifetime imaging microscopy (FLIM) measures variations in fluorescence decay times, providing spatial maps of molecular environments.

- **Organic LED Technologies**: Fluorescent and phosphorescent materials form the basis of organic LEDs (OLEDs), central in display technologies due to their efficiency and color versatility.

- **Nanotechnology Applications**: Incorporating photoluminescent properties into nanoparticles and quantum dots has opened avenues in drug delivery and diagnostics, enabling targeted therapeutic interventions coupled with tracking abilities.

Challenges in Photoluminescence Study

Researchers must navigate challenges such as photobleaching, where continued excitation results in irreversible fluorescence loss, and the need for environmentally benign, stable photoluminescent materials.

7.7 Photochemistry and Photophysical Processes

Photochemistry and photophysical processes are fundamental branches of chemistry that delve into the chemical and physical changes that occur upon the absorption of light. These disciplines play a key role in understanding the interactions of light with matter, which is vital for processes such as photosynthesis, vision, and the development of technological applications like solar energy conversion and fluorescence microscopy.

Fundamental Concepts

At the core of photochemistry is the absorption of photons by molecules, which leads to electronic excitations. These excitations cause the molecules to transition from a ground state to an excited electronic state, where they undergo various processes. Key concepts involving these processes include:

- **Photoexcitation:** Occurs when a molecule absorbs light energy, promoting an electron from a lower energy orbital to a higher energy orbital, creating an electronically excited state.

- **Relaxation Processes:** Once in an excited state, the molecule may release the absorbed energy through various pathways, restoring it to a lower energy state. This restoration may involve fluorescence, phosphorescence, internal conversion, or intersystem crossing.

- **Photochemical Reactions:** Chemical reactions initiated by the absorption of light, resulting in new chemical products. These reactions harness the energy of light to break and form chemical bonds, often entering pathways with distinct kinetics and mechanisms compared to thermal processes.

Photophysical Processes

Photophysical processes encompass changes that occur in molecules after they absorb light and transition to excited states without altering their chemical structure. They include:

- **Fluorescence:** As a molecule returns from an excited singlet state S_1 to its ground state S_0, light is emitted. This process is rapid, occurring on the nanosecond timescale.

- **Phosphorescence:** Involves the transition from an excited triplet state T_1 back to the ground state S_0. This process is much slower than fluorescence due to the involvement of a forbidden spin state transition.

- **Internal Conversion:** A non-radiative process where an excited molecule loses energy by converting it into vibrational energy within the same electronic state or transitioning from a higher to a lower electronic state without radiation.

- **Intersystem Crossing:** A process where molecules transition between electronic states of different spin multiplicities, notably from singlet to triplet states, facilitating phenomena like phosphorescence.

Photochemical Reactions

Photochemical reactions exploit photophysical processes to facilitate or drive chemical transformations:

- **Photosynthesis:** A quintessential photochemical reaction where plants, algae, and cyanobacteria convert light energy into chemical energy, producing glucose and oxygen by using sunlight to drive the conversion of carbon dioxide and water.

- **Vision:** Based on photochemical processes in light-sensitive cells in the retina where the chromophore retinal undergoes isomerization upon absorbing light, initiating a cascade that results in vision.

- **Photoisomerization:** An example is the cis-trans isomerization seen in compounds like azobenzene, where light can alter the spatial arrangement of atoms in a molecule, which is essential for molecular switches and optical data storage.

- **Photopolymerization:** Light initiates the polymerization of monomers, a process integral to the manufacture of plastics, resins, and advanced composite materials.

Examples and Applications

7.7. PHOTOCHEMISTRY AND PHOTOPHYSICAL PROCESSES

Photochemistry has revolutionized multiple domains by providing understanding and control over light-matter interactions:

- **Solar Energy Conversion:** Photochemical processes in photovoltaic cells convert sunlight into electricity, while photochemical reactions in photocatalysts convert sunlight into chemical energy, integral for sustainable energy solutions.

- **Fluorescence Microscopy:** Utilizes fluorescent dyes or proteins to illuminate cellular and molecular structures, enhancing biotechnological research and diagnostics.

- **Photocatalysis:** Photocatalysts, like titanium dioxide, facilitate chemical reactions upon absorbing light, offering pathways for degrading pollutants, splitting water to generate hydrogen, and organic transformations.

- **UV Protection:** Photochemistry underpins formulations of sunscreens which absorb or reflect harmful UV radiation, protecting biological tissues from damage.

Reactive Intermediates

Photochemistry often involves highly reactive intermediates like radical species, carbenes, or nitrenes that are harnessed in synthetic applications:

- **Radicals:** Play a role in polymerization and atmospheric chemistry. Radical formation can be tuned by modifying light exposure and molecular precursors, advancing methods in radical polymerization.

- **Photon-Initiated Reactions:** Facilitate organic transformations, expanding synthesis routes that are often inaccessible through thermal methods, with applications in industry and academia for creating complex organic compounds.

Challenges and Advances

Advancements in theoretical and computational chemistry alongside experimental techniques continually enhance photochemical studies:

- **Laser Spectroscopy:** High-resolution laser sources enable detailed investigations of ultrafast processes and mechanisms occurring on femtosecond to attosecond timescales, deepening our understanding of transition states and reaction pathways.

- **Computational Models:** Help simulate photophysical and photochemical processes, predicting behavior and guiding experimental designs.

- **Nanostructures in Photochemistry:** Use of nanomaterials, with unique optical properties derived from quantum confinement, is revolutionizing photochemical applications, from enhanced light harvesting to novel catalytic surfaces.

Photochemistry and photophysical processes remain crucial in driving innovations across chemistry, material science, and biotechnology. Continued research leads to the discovery of novel substances and methods, promising innovative solutions to contemporary energy challenges, medical diagnostics, and environmental sustainability. Through strategic manipulation of light and matter, these disciplines open a window into understanding and harnessing the quantum mechanical realm of chemical interactions.

Chapter 8

Surface Chemistry and Catalysis

Surface chemistry encompasses the study of phenomena occurring at interfaces, where the unique attributes of surfaces profoundly influence chemical reactions and material behaviors. This chapter delves into the principles of adsorption, surface tension, and the energetics of surface interactions, revealing how these factors dictate catalytic efficiency and selectivity. The exploration includes the mechanistic insights of heterogeneous catalysis, through which active sites transform reactant molecules, and extends to enzymatic and industrial applications harnessing catalytic principles. Through sophisticated surface characterization techniques, the understanding of molecular organization and reactivity at interfaces is refined, underscoring the critical role surface chemistry plays in innovations across diverse scientific and technological domains.

8.1 Fundamentals of Surface Chemistry

Surface chemistry is a pivotal field that explores the interactions occurring at the boundary between two phases. These phases can be any combination of solid, liquid, and gas, with the interactions influencing a range of physical and chemical properties. Central to this field are the phenomena of adsorption, surface tension, and the unique characteristics that define surfaces and interfaces.

Adsorption is a primary concept in surface chemistry, where molecules adhere to the surface of a solid or liquid. This process is categorized into physisorption (physical adsorption) and chemisorption (chemical adsorption). Physisorption arises due to van der Waals forces, which are relatively weak and result in reversible adsorption. In contrast, chemisorption involves the formation of chemical bonds between the adsorbate and the surface, leading to stronger and often irreversible adsorption.

The significance of adsorption lies in its applications across multiple scientific disciplines. It plays a vital role in catalysis, with adsorbate molecules interacting with the catalyst's surface to undergo reactions. In environmental science, adsorption is used for water purification and pollutant removal, where compounds adhere to activated carbon or similar materials. Moreover, adsorption techniques are utilized in chromatography, where it facilitates the separation of chemical mixtures.

Surface tension is another crucial concept in surface chemistry. It arises due to the cohesive forces between molecules at a liquid's surface, which are not counteracted by similar forces from above, leading to a "skin" effect. The magnitude of surface tension is influenced by the nature of the liquid and environmental conditions such as temperature.

Surface tension explains a range of phenomena, such as the ability of certain insects to walk on water and the formation of droplets. It also plays a role in various industrial applications, including the formulation of detergents and the stabilization of emulsions. Surface tension can be quantitatively measured by methods such as the capillary rise method and the drop weight method, each providing insights into the molecular interactions at the interface.

The characteristics of surfaces and interfaces are defined by their chemical composition and molecular arrangement, resulting in properties unique to the boundary itself rather than the bulk phases. Surface energy, a related concept, refers to the excess energy at the surface due to unsatisfied bonds when compared to the interior. Materials with high surface energy tend to be more reactive, making them important in the context of catalysis and adhesion.

When considering surfaces, roughness and texture are important attributes. Surface roughness can significantly affect properties such as wetting, friction, and optical behavior. Atomic and subatomic level details become significant

8.1. FUNDAMENTALS OF SURFACE CHEMISTRY

in nanomaterials, where the high surface-to-volume ratio dictates a substantial portion of material properties.

Exploring surface phenomena extends to understanding interfaces, which include the liquid-liquid interface observed in emulsions and the solid-solid interface critical in composite materials. Interfaces can exhibit distinct properties not found in the individual phases, lending themselves to the development of new materials with tailored characteristics.

Advanced techniques such as scanning electron microscopy (SEM) and atomic force microscopy (AFM) provide detailed insights into surface structure and texture. SEM offers high-resolution images by scanning the surface with a focused beam of electrons, revealing topology and composition. AFM provides a three-dimensional surface profile by measuring forces between a sharp probe and the surface at the atomic level. These techniques are indispensable tools for researchers aiming to elucidate surface characteristics and behaviors at the micro and nanoscale.

Beyond characterization, the controlled modification of surfaces allows for the enhancement of properties or the introduction of new functionalities. Techniques such as surface coating, functionalization, and deposition methods are employed in this regard. For instance, the application of a hydrophobic coating can impart water repellency to a surface, and specific chemical treatments can enable selective adsorption of target molecules.

The interplay of surface chemistry with materials science underpins developments in nanotechnology, where the modification and manipulation of surfaces at the nanoscale lead to groundbreaking applications. Surface chemistry principles are harnessed in developing drug delivery systems, sensors, and nanoelectronics, showcasing the interdisciplinary nature and broad impact of this field.

Overall, surface chemistry encompasses an extensive range of phenomena at the boundary between phases, playing an indispensable role in advancing both fundamental science and practical applications across diverse industries. Understanding these interactions is crucial for developing new materials and technologies, where surface properties are intentionally designed to meet specific requirements or achieve desired outcomes. As research in this field progresses, surface chemistry will continue to unlock new possibilities, driving innovation and addressing various scientific and technological challenges.

8.2 Adsorption Isotherms

Adsorption isotherms describe the relationship between the amount of adsorbate on the adsorbent and its concentration in the surrounding phase at constant temperature. They provide valuable insight into the nature of the adsorption process and the mechanics underlying surface interactions. Understanding adsorption isotherms is essential for catalysis, surface chemistry, and related fields, offering a foundation for designing processes and experiments.

One of the most fundamental adsorption isotherms is the Langmuir isotherm, proposed by Irving Langmuir in 1916. This model assumes that adsorption occurs at specific homogeneous sites within the adsorbent and that each site can hold only one molecule, leading to a monolayer adsorption. The model follows the assumptions of a uniform surface with no interactions between adsorbate molecules and reversible adsorption.

The Langmuir isotherm can be expressed mathematically as:

$$\theta = \frac{KP}{1+KP}$$

where θ is the fractional coverage of the surface, P is the pressure (or concentration for solutions) of the adsorbate, and K is the equilibrium constant related to the affinity between the adsorbate and the adsorbent.

The Langmuir isotherm is particularly useful when analyzing gases adsorbed on a solid surface. By transforming the equation, one can linearize the data and extract the Langmuir parameters. This facilitates determining the surface area of the adsorbent and provides insights into adsorption energies.

Despite its utility, the Langmuir model has limitations, especially in systems where the assumptions of homogeneity and monolayer adsorption do not hold. For such cases, alternative models are employed, one of which is the Freundlich isotherm.

Introduced by Herbert Freundlich in 1906, the Freundlich isotherm is an empirical model that describes adsorption on heterogeneous surfaces. Unlike the Langmuir model, it does not assume a finite number of adsorption sites and allows for multilayer adsorption. The Freundlich equation is given by:

$$q_e = K_F C_e^{1/n}$$

8.2. ADSORPTION ISOTHERMS

where q_e is the amount of adsorbate per unit mass of adsorbent, C_e is the equilibrium concentration of the adsorbate, K_F is the Freundlich constant indicative of adsorption capacity, and n is a heterogeneity factor.

The Freundlich model is particularly adept at modeling adsorption processes where the surface has varying affinities for the adsorbate, a scenario commonly encountered in the adsorption of dyes or organics on activated carbon and clays. It effectively captures the mechanisms in cases where adsorption occurs in pores or on surfaces with different energy distributions.

A further refinement comes from the BET (Brunauer–Emmett–Teller) isotherm, which extends the Langmuir model to multilayer adsorption. Developed by Stephen Brunauer, Paul Emmett, and Edward Teller in 1938, the BET isotherm is crucial for studying gas adsorption on solids at higher pressures. It incorporates the possibility of multiple layers of adsorption, relying on assumptions similar to Langmuir's but allowing for interactions between adsorbed molecules in the layer adjacent to the surface.

The BET equation takes the form:

$$\frac{P/(V(1 - P/P_0))}{P_0} = \frac{1}{V_m C} + \frac{(C-1)P}{V_m C P_0}$$

where V is the volume of gas adsorbed at pressure P, V_m is the monolayer adsorption capacity, P_0 is the saturation pressure, and C is the BET constant related to the energy difference between the first and higher layers.

BET isotherm models are widely used in characterizing the surface area and porosity of materials, making them indispensable in the development of catalysts and porous materials.

While these models form the backbone of adsorption isotherms, the complexity of real systems often leads to deviations that necessitate further refinement or the development of new models. The Temkin isotherm, for instance, recognizes the effects of indirect adsorbate/adsorbate interactions on adsorption isotherms and assumes that heat of adsorption decreases linearly with coverage.

Similarly, the Dubinin-Radushkevich (D-R) isotherm is often employed for adsorption studies of gases in microporous materials and provides insights into pore filling processes itself, assuming a Gaussian energy distribution.

For practical applications, isotherms help in designing adsorption systems for environmental remediation, such as the removal of pollutants from air or water. The selection of an appropriate adsorption model allows engineers to predict the performance and efficiency of adsorbent materials in real-world scenarios.

When interpreting adsorption data experimentally, it is crucial to match the data to an appropriate isotherm model. Parameters derived from these models aid in defining the operating conditions and the viability of different adsorbents. For instance, in catalysis, the surface area estimated from the BET isotherm is pivotal to determine the catalytic activity per unit area, a parameter essential for designing efficient catalysts.

Adsorption isotherms also intersect with biological systems, where the adsorption of molecules on cellular surfaces can affect biological activity. Analyzing such interactions through isotherms aids in understanding processes like enzyme-substrate binding and drug action mechanisms.

In summary, adsorption isotherms represent a critical element in surface chemistry, providing a quantitative method to examine how molecules interact with surfaces. Each model, with its assumptions and limitations, contributes to a comprehensive understanding of adsorption processes. As the discipline evolves, new isotherm models continue to emerge, addressing complex interactions and environmental conditions, thus maintaining the relevance and importance of adsorption isotherms in academic and industrial settings.

8.3 Surface Area and Porosity

Surface area and porosity are fundamental concepts in material science and surface chemistry, influencing the behavior and functionality of materials in applications such as catalysis, adsorption, and filtration. Their measurement and analysis are crucial for the development of advanced materials, including catalysts, adsorbents, and membranes.

Surface area is a measure of the total exposed area of a material and is typically expressed in square meters per gram (m^2/g). The surface area is particularly important in processes like catalysis and adsorption, where reactions or interactions occur at the surface. Porosity, on the other hand, refers to the fraction of the volume of voids over the total volume and is a crucial parameter in understanding the transport properties of gases and liquids within materials.

In porous materials, such as activated carbons, zeolites, and metal-organic frameworks (MOFs), the surface area is contributed significantly by the in-

8.3. SURFACE AREA AND POROSITY

ternal surfaces of the pores. The characterization of pore structure, including pore size, shape, and distribution, is essential for tailoring materials to specific applications. Porous materials with high surface areas and controlled pore sizes are desirable for catalysis, as they provide ample active sites for reactions and facilitate the diffusion of reactants and products.

The BET (Brunauer–Emmett–Teller) method is a widely used technique for measuring the specific surface area of materials. It builds upon the principles of gas adsorption and extends the Langmuir theory to multilayer adsorption. The BET surface area analysis involves the adsorption of a gas (commonly nitrogen) onto the material surface at liquid nitrogen temperature, resulting in an adsorption isotherm that can be analyzed to determine the surface area.

The BET equation is expressed as:

$$\frac{P/(V(1 - P/P_0))}{P_0} = \frac{1}{V_m C} + \frac{(C - 1)P}{V_m C P_0}$$

where V is the volume of gas adsorbed at pressure P, P_0 is the saturation pressure, V_m is the monolayer adsorbed gas volume, and C is the BET constant. By plotting the BET equation, known as the BET plot, one can determine the linear region to derive V_m, and subsequently calculate the surface area.

Understanding porosity involves several parameters, including pore volume, pore size distribution, and pore geometry. Pores are typically categorized into micropores (less than 2 nm), mesopores (2–50 nm), and macropores (greater than 50 nm) based on the IUPAC classification. The characterization of these pore sizes is crucial for applications like molecular sieving, where only molecules of specific sizes can pass through the pores.

Techniques such as mercury intrusion porosimetry, gas adsorption, and electron microscopy are some of the common methods for assessing porosity. Mercury intrusion porosimetry involves forcing mercury into the pores under pressure and is particularly effective for detecting macroporosity. Gas adsorption methods, using gases like nitrogen or argon at various temperatures and pressures, enable the determination of microporosity and mesoporosity. Pore size distribution can also be analyzed by the BJH (Barrett-Joyner-Halenda) method, which is applied to desorption isotherms to provide information about mesopore sizes.

Recent advances in imaging techniques, such as X-ray computed tomography (XCT) and scanning electron microscopy (SEM), provide detailed visualizations of the pore structures. XCT, especially, offers a non-destructive means of obtaining three-dimensional representations of the internal pore networks, aiding in the analysis of complex porous systems.

The manipulation of surface area and porosity is central to designing materials for specific purposes. For instance, catalysts with high surface areas and tailored pore structures enhance reaction rates by providing increased accessibility to active sites and optimizing the transport of reactants and products. In adsorption processes, materials with specific pore size distributions target the selective removal of contaminants, as seen in gas separation and water purification.

In environmental applications, porous materials such as biochar and activated carbon are utilized for removing organic pollutants and heavy metals from water, exploiting their high surface areas and porous structures. Similarly, in gas storage, materials like MOFs and zeolites, with their defined pore networks, enable the efficient storage and release of gases such as hydrogen and methane.

In biotechnology, porous scaffolds are designed to mimic the extracellular matrix, supporting cell attachment and proliferation for tissue engineering. The control of porosity, in this case, is vital for nutrient flow and cellular migration, ensuring the successful creation of functional tissue constructs.

The role of surface area and porosity extends to energy applications, particularly in the development of battery electrodes and supercapacitors. High surface area materials with controlled porosity provide greater electrode/electrolyte contact areas, enhancing ionic transport and storage capacity.

Surface area and porosity also influence mechanical properties, where porous materials, such as foams and aerogels, exhibit unique mechanical characteristics like lightweight and high compression resistance, suitable for thermal insulation and impact absorption.

Overall, the meticulous characterization and engineering of surface area and porosity are instrumental in advancing a wide spectrum of technologies, from catalysis to energy storage and environmental remediation. As materials science progresses, the ability to tailor these properties at the nanoscale continues to expand the applications and performance of porous materials, underscoring their importance in scientific research and industrial development.

8.4 Catalysis and Catalytic Mechanisms

Catalysis is a fundamental process in chemistry that accelerates chemical reactions without undergoing permanent changes itself. It is a cornerstone in industrial applications, environmental protection, and biochemical processes. Understanding catalytic mechanisms and the distinctions between different types of catalysis is essential for the development and optimization of catalysts that improve reaction efficiency and selectivity.

Catalysts function by providing an alternative reaction pathway with a lower activation energy compared to the uncatalyzed reaction. This increased reaction rate is achieved through interactions with reactants, leading to the formation of intermediates that facilitate the transformation to products. The effectiveness of a catalyst is determined by its activity, selectivity, and stability, which are influenced by the catalyst's physical and chemical properties.

Catalysis can be broadly divided into homogeneous and heterogeneous catalysis. Homogeneous catalysis involves catalysts that are in the same phase as the reactants, typically in a liquid solution. This type of catalysis is characterized by high selectivity and uniform molecular interactions. A classic example is the acid-catalyzed esterification of carboxylic acids with alcohols, where a proton acts as the catalyst in a homogeneous medium.

The catalytic cycle of homogeneous catalysis often involves coordinated metal complexes or organometallic compounds that undergo changes in oxidation state or coordination geometry during the reaction. Mechanistically, these processes can include steps of coordination, oxidative addition, reductive elimination, and ligand exchange, as observed in olefin polymerization catalyzed by metallocenes.

One advantage of homogeneous catalysts is their ability to provide well-defined active sites, leading to high specificity. However, separating the catalyst from products can be challenging, which has spurred interest in developing methods for catalyst recovery and recycling.

In contrast, heterogeneous catalysis involves catalysts that are in a different phase than the reactants, typically solid catalysts used in gas or liquid phase reactions. This type, exemplified by solid catalysts that interact with gaseous or liquid reactants, finds immense application in industrial processes due to the ease of catalyst separation and regeneration.

The mechanisms underlying heterogeneous catalysis often involve the adsorption of reactants on the catalyst surface, followed by a series of surface reactions and desorption of products. Key steps include the adsorption of reactants, surface diffusion, reaction at active sites, and desorption of products. The effectiveness of a heterogeneous catalyst relies heavily on its surface properties, such as surface area, morphology, and the nature of active sites.

A widely studied example of heterogeneous catalysis is the Haber-Bosch process, which synthesizes ammonia from nitrogen and hydrogen gases over iron-based catalysts. In this process, nitrogen molecules adsorb on the catalyst surface and dissociate into nitrogen atoms, which then react with hydrogen to form ammonia. The high activation energy for nitrogen dissociation is overcome by the catalyst, emphasizing the critical role of catalysts in facilitating reactions that are otherwise too slow to be commercially viable.

Catalytic mechanisms at the molecular level can vary greatly depending on the type of reaction and the nature of the catalyst. In enzyme catalysis, for example, the substrate binds to the active site of the enzyme, forming an enzyme-substrate complex that stabilizes the transition state and lowers the activation energy. Enzyme catalysis is often characterized by exquisite specificity and efficiency, guided by mechanisms such as covalent catalysis, acid-base catalysis, and metal ion catalysis.

Transition state theory provides a framework for understanding catalytic mechanisms, linking the structure of the transition state to the rate constant of the reaction. Catalysts stabilize the transition state, effectively reducing the energy barrier that reactants must overcome. This conceptual understanding is invaluable in rational catalyst design, where modifications to catalyst structure aim to enhance interaction with the transition state.

The development of new catalysts and catalytic systems is driven by a multitude of factors, including economic considerations, environmental impact, and the pursuit of sustainable chemical processes. For instance, the field of green chemistry strives to create catalysts that minimize waste and energy consumption, using resources more efficiently.

Recent advances in catalysis research have focused on the integration of nanotechnology and catalysis. Nanocatalysts, with their high surface area-to-volume ratios and tunable electronic properties, offer enhanced activity and selectivity. Their nanoscale dimensions allow for engineered surface features and active site architectures, facilitating unique catalytic behaviors.

Furthermore, computational chemistry plays a significant role in exploring catalytic mechanisms and the design of novel catalysts. Density functional theory (DFT) calculations provide insights into the electronic structure of catalysts and reaction intermediates, aiding in the understanding of catalytic pathways and the prediction of catalyst performance.

The development of bifunctional and multifunctional catalysts, which combine different catalytic activities within a single material, represents another frontier in catalysis. These catalysts are capable of orchestrating complex reaction networks, such as tandem reactions, where multiple transformations occur sequentially or simultaneously under controlled conditions.

In industrial settings, catalysis is integral to the production of chemicals, fuels, and polymers, with advances continuing to improve the efficiency and sustainability of these processes. Emerging applications, such as the catalytic conversion of carbon dioxide to value-added products and the hydrogenation of carbon resources, exemplify the pivotal role of catalysis in addressing global challenges related to energy and the environment.

In summary, catalysis and catalytic mechanisms form the backbone of modern chemical sciences, with broad implications across an array of fields. By lowering the activation energy and providing alternative pathways for chemical reactions, catalysts facilitate processes that are essential to human life and technological advancement. The continuous exploration and understanding of catalytic mechanisms drive the innovation of more efficient, selective, and sustainable catalysts, contributing significantly to solving contemporary scientific and industrial challenges.

8.5 Enzyme Catalysis and Industrial Applications

Enzyme catalysis represents a remarkable branch of catalysis, where biological molecules, primarily proteins, accelerate chemical reactions with extraordinary specificity and efficiency. Enzymes are nature's catalysts, enabling complex biochemical processes crucial for sustaining life. They operate under mild conditions of temperature and pressure, creating opportunities for environmentally friendly and sustainable industrial processes.

The fundamental mechanism of enzyme catalysis involves the formation of an enzyme-substrate complex. The active site of an enzyme, typically a small pocket or groove on its surface, is designed to bind specific substrate molecules. This specificity is attributed to the unique three-dimensional conformation of the enzyme, allowing precise interactions, often described by the "lock and key" model or the more flexible "induced fit" model.

Upon binding the substrate, enzymes stabilize the transition state of the reaction, thereby lowering the activation energy required for the conversion of substrate to product. This process may involve several catalytic strategies, such as covalent catalysis, where transient covalent bonds form between the enzyme and the substrate; acid-base catalysis, where functional groups within the enzyme act as proton donors or acceptors to facilitate the reaction; and metal ion catalysis, which involves metal cofactors that aid in substrate orientation and electron transfer.

One classic example of enzyme catalysis is the action of the enzyme lysozyme. Lysozyme catalyzes the hydrolysis of $\beta(1 \rightarrow 4)$ glycosidic bonds in peptidoglycan, a structural component of bacterial cell walls. Through acid-base catalysis and strain induction, lysozyme effectively disrupts the integrity of bacterial cells.

In recent years, the utilization of enzymes in industrial applications has expanded dramatically, primarily due to advancements in biotechnology and a growing emphasis on sustainable processes. Enzymes are now integral to the production of pharmaceuticals, food and beverages, biofuels, and environmentally friendly cleaning products.

In the pharmaceutical industry, enzymes are employed in the synthesis of chiral drug molecules, which require high selectivity. Enantioselective reactions catalyzed by enzymes, such as lipases and esterases, enable the production of enantiomerically pure compounds, critical for drug efficacy and safety. The synthesis of the semi-synthetic antibiotic amoxicillin, an example, involves the use of enzymes to ensure the chirality of its active pharmaceutical ingredient.

The food and beverage industry extensively utilizes enzymes to enhance flavor, texture, and nutritional value. Amylases, for instance, break down starches into sugars, facilitating processes like bread making and brewing. Proteases improve dough properties in baking, while lactase is used to produce lactose-free dairy products, catering to those with lactose intolerance.

Enzyme technology also underpins the production of biofuels, an area of significant interest for renewable energy. Enzymes such as cellulases and hemicellulases break down plant biomass into fermentable sugars, which are then converted into ethanol by microbial fermentation. This conversion process is

8.5. ENZYME CATALYSIS AND INDUSTRIAL APPLICATIONS

central to the production of second-generation biofuels, which use non-food biomass sources, minimizing competition with food crops.

Environmental applications of enzymes extend to bioremediation, where enzymes degrade pollutants in soils and water, providing a green solution for environmental cleanup. Laccases, for example, catalyze the oxidation of phenolic and non-phenolic lignin-related compounds, contributing to the degradation of pollutants in wastewater treatment processes.

In industrial cleaning products, enzymes such as proteases, lipases, and amylases are incorporated to enhance the cleaning efficacy at lower temperatures, reducing the energy consumption and environmental footprint of these products.

Despite their numerous benefits, the industrial application of enzymes faces challenges, particularly related to stability and cost. Enzymes are sensitive to denaturation and loss of activity outside their optimal conditions, such as extreme temperatures, pH, or presence of organic solvents. To overcome these limitations, strategies such as enzyme immobilization, protein engineering, and the use of extremophiles (organisms that thrive in extreme environments) are employed to enhance enzyme stability and performance.

Enzyme immobilization involves attaching enzymes to solid supports, enabling easy separation and reuse, thereby enhancing their stability and reducing costs. Immobilized enzymes also allow for continuous processes in industrial settings, improving efficiency and productivity.

Protein engineering techniques, including directed evolution and rational design, are instrumental in optimizing enzyme properties. Directed evolution mimics the natural evolutionary process through iterative rounds of mutation and selection, yielding enzymes with enhanced stability, activity, or altered substrate specificity. Rational design, driven by computational modeling and structural biology insights, enables precise modifications to an enzyme's active site or structural framework, tailoring its function for a specific industrial application.

The exploration of extremozymes, enzymes derived from extremophiles, offers another avenue for extending the operational range of enzymes under industrial conditions. These enzymes are inherently stable at high temperatures, extreme pH levels, or high salinity, making them ideal candidates for industrial processes that operate under harsh conditions.

Regulatory and economic considerations also play a role in the utilization of enzymes in industry. The safety, efficacy, and environmental impact of enzyme-based processes are subject to rigorous assessment, aligning with sustainable development goals and reducing the carbon footprint of industrial production.

In summary, enzymes are formidable catalysts with diverse industrial applications, powered by their specificity and efficiency under mild conditions. Advances in biotechnology continue to expand the possibilities for enzyme catalysis, creating innovative and sustainable solutions across a broad range of industries. As the focus on green chemistry and sustainability intensifies, the role of enzyme catalysis in shaping the future of industrial processes becomes increasingly pivotal, driving advancements in efficiency, selectivity, and environmental stewardship.

8.6 Surface Characterization Techniques

Surface characterization is an essential aspect of surface science and materials engineering, providing detailed insights into the physical, chemical, and mechanical properties of surfaces and interfaces. Understanding these properties is critical for a wide range of applications, including catalysis, nanotechnology, corrosion prevention, and biomaterials development. Various techniques have been developed to probe surface characteristics at macroscopic, microscopic, and even atomic levels.

- **Scanning Electron Microscopy (SEM)** is a widely used technique to image and analyze the surface morphology and composition of materials with high resolution. SEM works by scanning a focused electron beam over a specimen, which interacts with the atoms at the surface, producing signals in the form of secondary electrons, backscattered electrons, and characteristic X-rays. These signals are collected to form detailed images of the surface topology and to identify compositional elements via energy-dispersive X-ray spectroscopy (EDS).

- The resolution of SEM can reach nanometer levels, making it an invaluable tool for characterizing surface structures and defects. For instance, in electronics, SEM helps in analyzing the surface features of integrated circuits and identifying failures and sub-micron defects. In materials science, SEM characterizes coating thickness, surface roughness, and

8.6. SURFACE CHARACTERIZATION TECHNIQUES

fracture surfaces, providing insights into material behavior under different conditions.

- **Transmission Electron Microscopy (TEM)** offers an alternative approach to surface characterization, capable of providing information about the internal and surface structure of materials at the atomic level. Unlike SEM, where electrons are reflected, TEM involves transmitting electrons through an ultra-thin sample. The interaction of electrons with the sample creates an image that can reveal lattice structures and defects.

- TEM is particularly advantageous for analyzing nanomaterials, crystallography, and interfaces within layered materials. High-resolution TEM (HRTEM) analysis can elucidate atomic arrangements in nanoparticles, semiconductor devices, and biomaterials, aiding in understanding materials' properties and behaviors.

- **Atomic Force Microscopy (AFM)** is another crucial technique for characterizing surfaces with high precision. AFM operates by scanning a sharp tip over the specimen surface, measuring the forces between the tip and the surface to generate a topographic map. Unlike SEM and TEM, which rely on electron beams, AFM provides true three-dimensional surface profiles without the need for coating or conductive surfaces.

- AFM is particularly useful for analyzing surfaces at the nanometer scale, such as thin films, polymers, and biological materials. It can quantify surface roughness, measure mechanical properties such as hardness and adhesion, and provide insights into electrical and magnetic surface properties. Its application in nanotechnology is extensive, enabling the characterization of nanostructures, nanocomposites, and soft matter systems with unparalleled detail.

- **Surface characterization** extends beyond microscopic techniques, encompassing spectroscopic methods that yield chemical information about surfaces. X-ray Photoelectron Spectroscopy (XPS), also known as Electron Spectroscopy for Chemical Analysis (ESCA), is a powerful tool for determining the elemental composition and chemical state of atoms at the surface. XPS involves irradiating a material with X-rays and measuring the kinetic energy of ejected photoelectrons, providing quantitative chemical analyses with a depth resolution of a few nanometers.

- XPS is invaluable for understanding surface chemistry, particularly in the study of oxides, corrosion layers, and surface modifications. It helps determine oxidation states, identify functional groups, and quantify surface contaminants, which is essential for applications in corrosion science, semiconductor manufacturing, and surface coating development.

- **Auger Electron Spectroscopy (AES)** is another surface-sensitive technique that provides information about the elemental composition. AES uses a focused electron beam to excite atoms on the surface, causing ejection of Auger electrons whose energies are characteristic of elements present. AES is highly sensitive to surface layers and thin films, widely applied for characterizing catalysts, surface treatments, and microelectronics.

- To complement spectroscopic techniques, **Secondary Ion Mass Spectrometry (SIMS)** offers a method to analyze surface composition using ion sputtering. SIMS provides high sensitivity for detecting trace elements and isotopes, generating mass spectra that can reveal surface and near-surface composition. Its applications range from studying semiconductor interfaces to analyzing geological samples and thin film coatings.

- For more comprehensive analysis, **Dual-Energy X-ray Absorptiometry (DXA)** is used to assess the density and composition of surface layers. It is often used in combination with other spectroscopic techniques to gain a deeper understanding of material properties, especially in research involving complex materials and biomimetic surfaces.

- In addition to analytical techniques, surface characterization also relies on computational methods such as **Molecular Dynamics (MD)** and **Density Functional Theory (DFT)** simulations. These approaches model surface interactions and predict behaviors under various conditions, providing insights into surface reactivity, adsorption processes, and catalytic mechanisms.

The integration of surface characterization techniques enables a holistic view of surface properties and behaviors, essential for advancing material design and innovation. For example, in biomaterials development, characterizing surface chemistry and topography informs bio-compatibility and guides the design of implants and tissue scaffolds.

The development of energy materials, such as fuel cells and solar cells, also relies on precise surface characterization to optimize efficiency and performance. By studying the surface properties of catalytic materials, researchers can tailor surface features to enhance catalytic activity and stability, driving advancements in sustainable energy technologies.

In corrosion science, understanding how materials interact with their environment at the surface level is paramount for developing effective corrosion-resistant coatings and treatments. Surface characterization techniques enable the analysis of protective oxide layers, providing insights necessary for prolonging material lifespan and ensuring structural integrity.

Surface characterization techniques encompass a diverse array of microscopic, spectroscopic, and computational approaches, each offering unique insights into the complex world of surfaces and interfaces. By providing detailed information on structure, composition, and properties, these techniques underpin innovations across industries, from electronics and energy to healthcare and environmental science, driving the development and optimization of advanced materials and technologies.

8.7 Applications of Surface Chemistry in Nanotechnology

Surface chemistry plays a pivotal role in nanotechnology, where the unique properties of materials emerge at the nanoscale due to a significant increase in the surface area-to-volume ratio. At this scale, surface phenomena dominate the material's behavior, leading to novel applications across various fields, including medicine, electronics, energy, and environmental science. Understanding and manipulating surface interactions enable the design and development of nanostructured materials with tailored functionalities.

One of the hallmark applications of surface chemistry in nanotechnology is in the field of drug delivery systems. Nanoparticles serve as carriers for therapeutic agents, allowing for controlled and targeted delivery. Surface modification of these nanoparticles with polyethylene glycol (PEGylation), targeting ligands, or biomolecules enhances their stability, biocompatibility, and specificity. For instance, modifying the surface of nanoparticles with antibodies or peptides enables targeted delivery to cancer cells, minimizing systemic toxicity and improving therapeutic efficacy.

Surface chemistry is also crucial in the fabrication of biosensors, which are devices that convert a biological response into an electrical signal. Nanomaterials such as gold nanoparticles, carbon nanotubes, and graphene serve as transducers in these sensors due to their high surface area, conductivity, and ability to immobilize biomolecules. Surface functionalization with specific recognition elements, such as enzymes, antibodies, or nucleic acids, enhances the sensor's selectivity and sensitivity, enabling the detection of various analytes, including glucose, pathogens, and pollutants.

In electronics, surface chemistry facilitates the development of nanoscale components, such as transistors, capacitors, and interconnects. The miniaturization of electronic devices demands precise control over the surface properties of materials. For instance, the functionalization of silicon surfaces with organic molecules can modulate the electronic properties, enabling the integration of molecular electronics. Similarly, the passivation of semiconductor surfaces is essential to prevent oxidation and improve device performance.

Surface chemistry is integral to the field of catalysis, particularly in the design of nanocatalysts with enhanced activity and selectivity. Nanoparticles such as platinum, palladium, and gold exhibit unique catalytic properties due to their high surface area and active sites. Surface modification of these nanoparticles with ligands or capping agents can tune their electronic properties and steric environments, optimizing the catalytic process. For example, in hydrogenation reactions, the modification of palladium nanoparticles with dendrimer ligands enhances selectivity and reaction rates.

The development of energy storage and conversion devices, including batteries, fuel cells, and supercapacitors, benefits significantly from surface chemistry. Nanostructured materials with controlled surface properties improve the electrochemical performance of electrodes. Surface coatings on lithium-ion battery electrodes, for instance, enhance their stability and cycling performance. In fuel cells, the design of nanostructured catalysts with active and stable surfaces is crucial for efficient oxygen reduction and hydrogen oxidation reactions.

Photovoltaic devices, particularly those based on quantum dots and dyes, leverage surface chemistry for improved light harvesting and conversion efficiency. The functionalization of quantum dots with organic molecules facilitates charge transfer, while surface passivation reduces recombination losses. In dye-sensitized solar cells, the anchoring of dye molecules onto semiconductor surfaces is fundamental for light absorption and electron injection.

8.7. APPLICATIONS OF SURFACE CHEMISTRY IN NANOTECHNOLOGY

Environmental applications of nanotechnology are profoundly impacted by surface chemistry, particularly in the context of water purification and air quality improvement. Nanomaterials such as titanium dioxide and silver nanoparticles are employed for their photocatalytic and antimicrobial properties. Surface modification of these materials enhances their activity, stability, and selectivity for degrading pollutants or inactivating pathogens.

Nanotechnology also finds applications in the development of advanced coatings and paints, where surface chemistry determines properties such as hydrophobicity, oleophobicity, and antifouling capabilities. Coatings with nanoscale features provide durable protection and self-cleaning properties, reducing maintenance costs and extending material lifespan.

In textiles, surface chemistry enables the incorporation of nanoscale additives that impart unique functionalities, such as water and stain resistance, UV protection, and antimicrobial activity. For instance, nano-coatings on fabrics can repel liquids and resist staining, enhancing the durability and utility of clothing and upholstery.

Despite the advances in nanotechnology, challenges remain, particularly concerning the environmental, health, and safety implications of nanomaterials. Surface chemistry plays a role in addressing these challenges by enabling the design of safer and more sustainable nanomaterials. Surface modifications can improve the dispersion and stability of nanomaterials in biological and environmental media, reducing the risk of aggregation and toxicity. Additionally, surface chemistry can facilitate the degradation or recycling of nanomaterials, minimizing their environmental impact.

In sum, surface chemistry is foundational to the advancement of nanotechnology across diverse application areas. By providing tools for the precise manipulation of nanoscale materials, surface chemistry underpins innovations that address contemporary challenges in health, energy, environment, and industry. As the field continues to evolve, the integration of surface chemistry with nanotechnology promises to drive further breakthroughs and foster the development of next-generation materials and devices.

Chapter 9

Electrochemistry and Conductivity

Electrochemistry explores the interplay between electrical energy and chemical change, offering invaluable insights into processes ranging from battery operation to metal corrosion. This chapter examines the fundamental principles governing electrochemical cells, elucidating how redox reactions generate electric currents and vice versa. The quantitative frameworks provided by the Nernst equation and conductometric techniques are explored, revealing the dependence of electrochemical potential and conductivity on environmental conditions. By investigating galvanic and electrolytic cells, the theoretical principles are connected to practical applications in energy storage, electroplating, and biosensing. Thus, electrochemistry serves as a cornerstone for understanding and harnessing the intersection of chemical and electrical phenomena.

9.1 Fundamentals of Electrochemistry

Electrochemistry encompasses the study of chemical processes that facilitate the transfer of electrons. At the core of these processes are redox reactions, electrodes, and electrochemical cells, which together form the underlying structure of electrochemical systems. This section discusses the fundamental concepts, offering detailed insights into redox reactions, the structure and function of electrodes, and the operation and components of electrochemical cells.

Redox reactions, or oxidation-reduction reactions, serve as the foundation of electrochemical processes. In these reactions, electrons are transferred between species, with one species undergoing oxidation and losing electrons, while the other undergoes reduction and gains electrons. The overall reaction can be split into two half-reactions, each representing either oxidation or reduction. Consider the classic example of the zinc-copper electrochemical cell: zinc metal loses electrons to form zinc ions

$$Zn \longrightarrow Zn^{2+} + 2\,e^- \qquad (9.1)$$

while the copper ions gain electrons to form copper metal

$$Cu^{2+} + 2\,e^- \longrightarrow Cu \qquad (9.2)$$

Here, zinc is oxidized, serving as the reducing agent, and copper is reduced, acting as the oxidizing agent.

The understanding of electrodes is crucial in electrochemistry, as they are the mediums through which electrons are transferred in and out of a chemical system. Electrodes are classified into two main types: anodes and cathodes. In a galvanic cell, the anode is the site of oxidation and consequently is negatively charged, while the cathode is the site of reduction and is positively charged. In electrolytic cells, the polarities are reversed because external voltage drives the reactions.

Electrochemical cells are devices that convert chemical energy into electrical energy or vice versa, facilitating the flow of electrons through an external circuit. These cells are categorized primarily into galvanic cells, which produce electrical energy from spontaneous chemical reactions, and electrolytic cells, which consume electrical energy to drive non-spontaneous reactions. Each electrochemical cell consists of two half-cells connected by a salt bridge or a porous membrane allowing ions to flow between them, thereby maintaining charge neutrality.

The salt bridge, or sometimes a porous disk in simpler setups, is an essential component, as it allows the migration of ions without permitting the mixing of

9.1. FUNDAMENTALS OF ELECTROCHEMISTRY

different solution compartments. This migration sustains the electrochemical reaction by balancing the charges in both the anode and cathode compartments, preventing the buildup of charge that would otherwise halt the electron flow.

In-depth understanding of electrode potentials is fundamental to manipulating electrochemical reactions mathematically and practically. The standard electrode potential, E^0, is a measure of the tendency of a chemical species to be reduced, measured under standard conditions (25°C, 1 M concentration, and 1 atm pressure). Reference electrodes, such as the standard hydrogen electrode (SHE), are used to measure E^0 values. The SHE is defined to have a potential of 0.00 V, serving as a neutral benchmark for comparison between different electrodes.

The electrochemical series, an arrangement of elements in the order of their standard electrode potentials, provides a reference for predicting the direction of redox reactions and understanding the relative reactivity of different chemical species. For instance, elements with higher (more positive) standard electrode potentials are stronger oxidizing agents compared to those with lower potentials. An illustrative application is predicting the feasibility of a redox reaction using the potential differences derived from the electrochemical series.

One of the pivotal concepts in electrochemistry is the electromotive force (EMF) of a cell, which quantifies its ability to drive an electric current. The EMF is calculated by taking the difference between the standard reduction potentials of the cathode and anode:

$$\text{EMF} = E_c^0 - E_a^0$$

where E_c^0 and E_a^0 are the standard electrode potentials for the cathode and anode, respectively. A positive EMF signifies a spontaneous reaction under standard conditions, thereby providing the theoretical maximum voltage the cell can deliver.

Further, the kinetics of electron transfer and mass transport play crucial roles in the efficiency of electrochemical cells. The electron transfer rate is dictated by the reaction kinetics at the electrode interface, while the dynamics of diffusion, migration, and convection influence how reactants reach the electrode surface. Understanding these phenomena requires both qualitative and quantitative analyses to optimize the design and performance of electrochemical devices.

In practical applications, electrochemical cells form the basis for numerous technologies, including batteries, fuel cells, and electroplating systems. The versatility of electrochemical cells lies in their ability to use different materials and configurations to achieve desired outcomes. For example, in batteries such as Lithium-ion, different electrode materials are chosen to maximize charge and energy density. Fuel cells, on the other hand, harness electrochemical reactions of fuels such as hydrogen to generate electricity efficiently and sustainably.

The interdisciplinary nature of electrochemistry extends its implications beyond pure chemistry, impacting fields like materials science, environmental science, and industrial processes. In corrosion, for instance, the principles of electrochemical cells help understand and control the electrochemical degradation of metals. In environmental monitoring, electrochemical sensors provide tools for detecting pollutants with high sensitivity and selectivity. Industrial electroplating exploits controlled electrochemical deposition to coat substrates with thin layers of material, enhancing qualities such as conductivity, corrosion resistance, and aesthetic appeal.

Recent advancements in computational electrochemistry have furthered our understanding by simulating complex electrochemical phenomena at atomic and molecular levels. Such simulations enable the detailed investigation of reaction mechanisms, electrode-electrolyte interfaces, and transient behaviors under various conditions, lending insights that drive innovation in material design and process optimization.

Exploring the depths of electrochemistry involves pursuing both the theoretical and practical facets of electrochemical systems. Understanding the movement of ions and the behavior of electrodes under different conditions allows for the development of more efficient, sustainable, and resilient technologies. As we delve into electrochemical systems, the combination of empirical data and theoretical frameworks paves the way for advancements that meet the growing demands of a technology-driven society.

9.2 Nernst Equation and Electrochemical Potentials

The Nernst equation is one of the cornerstone mathematical relationships in electrochemistry that allows us to calculate the cell potential under non-

9.2. NERNST EQUATION AND ELECTROCHEMICAL POTENTIALS

standard conditions. This tool is indispensable for understanding electrochemical equilibrium and provides insights into how ion concentration gradients affect the potential of electrochemical cells. In this section, we explore the derivation and implications of the Nernst equation, along with its application in predicting the behavior of cells in varied conditions.

Initially, the study of electrochemical cells under standard conditions involves the use of standard electrode potentials, E^0, to predict the voltage of a cell. However, real-world systems often operate under non-standard conditions, where concentrations, pressures, and temperatures differ from those in the standard state. The Nernst equation bridges this gap, allowing for the determination of the cell potential by incorporating these practical variations in state.

The Nernst equation is derived from the fundamental principles of chemical thermodynamics. At the equilibrium state, the change in Gibbs free energy (ΔG) is related to the cell potential (EMF) and the number of moles of electrons transferred (n), expressed by the equation:

$$\Delta G = -nF\text{EMF}$$

where F is Faraday's constant (96,485 C/mol). Under standard conditions, the change in Gibbs free energy can also be related to the equilibrium constant (K) and is given by:

$$\Delta G^0 = -RT \ln K$$

where R is the universal gas constant (8.314 J/mol K) and T is the temperature in Kelvin. The Nernst equation results from combining these thermodynamic expressions and addressing non-standard state conditions:

$$\text{EMF} = E^0 - \frac{RT}{nF} \ln Q$$

where Q is the reaction quotient, representing the relative concentrations of products to reactants at any point in the reaction:

$$Q = \frac{\text{activities of products}}{\text{activities of reactants}}$$

The Nernst equation allows us to evaluate the potential of an electrochemical cell based on the differences in ion concentrations between the two half-cells.

For reactions at 25°C (298 K), the equation can be expressed in a simpler form by substituting numerical values for R and F, leading to:

$$\text{EMF} = E^0 - \frac{0.0592}{n} \log Q$$

In the case of concentration cells, where the electrodes are constructed from the same material but immersed in solutions of different concentrations, the Nernst equation clearly illustrates how potential difference is generated solely from the concentration gradient.

Consider a concentration cell with a specific half-cell reaction:

$$M^{n+} + ne^- \longrightarrow M$$

where the two half-cells contain the same metal and associated ion, but with varying ion concentrations $[M^{n+}]_{\text{cathode}}$ and $[M^{n+}]_{\text{anode}}$. The Nernst equation for such a cell becomes:

$$\text{EMF} = \frac{0.0592}{n} \log \left(\frac{[M^{n+}]_{\text{cathode}}}{[M^{n+}]_{\text{anode}}} \right)$$

This shows that the EMF is directly related to the logarithmic ratio of the ion concentrations, providing a concise way to measure concentration effects on potential in chemical sensors and selectivity in ion transport applications.

Electrochemical equilibrium is another domain where the Nernst equation plays a critical role. At equilibrium, the cell potential (EMF) becomes zero, and the reaction quotient (Q) matches the equilibrium constant (K). The Nernst equation under these conditions reveals the relationship between standard potentials and equilibrium concentrations.

In physiological systems, the Nernst equation is crucial for understanding membrane potentials. Biological cells often maintain different concentrations of ions across cell membranes, creating potential differences that are fundamental to processes such as nerve impulse transmission. The Nernst equation calculates the electrochemical potential for an ion across a membrane with differing intracellular and extracellular concentrations:

$$E_{\text{ion}} = \frac{RT}{zF} \ln \left(\frac{[\text{ion}]_{\text{outside}}}{[\text{ion}]_{\text{inside}}} \right)$$

where z is the charge number of the ion. For monovalent ions at body temperature ($\sim 37°C$ or 310 K), the equation simplifies, due to physiological scaling and constants, to:

$$E_{\text{ion}} = \frac{61.5}{z} \log \left(\frac{[\text{ion}]_{\text{outside}}}{[\text{ion}]_{\text{inside}}} \right)$$

The predictive power of the Nernst equation supports the design of electrochemical devices such as batteries. For example, in a lithium-ion battery, the battery's voltage is contingent on the concentrations of lithium ions within the anode and cathode compartments. By using the Nernst equation, manufacturers can optimize the concentrations to enhance performance and lifespan.

In corrosion science, the Nernst equation is employed to understand the potentials at which metals corrode or remain stable in various environments. For a metal M in an oxidizing environment, the corrosion potential is derived by analyzing the equilibrium between oxidized and reduced species, as influenced by ambient conditions.

Moreover, the application of the Nernst equation extends to electrochemical titrations, providing a basis for accurately determining the endpoint of titrations involving redox reactions. By plotting the potential against the volume of titrant added, a sharp change in potential can signify the endpoint. The equation thus plays a crucial role in analytical chemistry, enhancing methods for titration beyond traditional visual indicators.

The Nernst equation elucidates the response of electrochemical cells to variations in systemic conditions, bridging theoretical predictions with empirical phenomena. Its breadth encapsulates a framework applicable across a versatile array of scientific and engineering challenges, confirming it as a fundamental tool for electrochemical exploration and innovation. With the Nernst equation, the quantitative description of potential changes becomes a gateway to optimizing and controlling the myriad applications inherent in electrochemical systems.

9.3 Galvanic and Electrolytic Cells

Electrochemical cells are fundamental devices that facilitate the interconversion between chemical energy and electrical energy. Among these, galvanic (or voltaic) cells and electrolytic cells represent two major types, each with distinct operational principles and applications. This section delves into the construction, operation, and applications of both galvanic and electrolytic cells, providing comprehensive insights into their roles within electrochemistry.

Galvanic cells are devices that generate electrical energy from spontaneous redox reactions occurring within the cell. They are named after Luigi Galvani, an early researcher in bioelectricity, whose work laid the foundation for this study. The operation of a galvanic cell is rooted in spontaneous chemical reactions, whereby the free energy derived from redox reactions drives the movement of electrons through an external circuit.

A classic example of a galvanic cell is the Daniell cell, composed of a zinc half-cell and a copper half-cell. In this setup, a zinc electrode immersed in a zinc sulfate solution serves as the anode, while a copper electrode in a copper sulfate solution operates as the cathode. The half-reactions can be described as:

$$\text{Anode (oxidation): } Zn\,(s) \longrightarrow Zn^{2+}\,(aq) + 2\,e^-$$

$$\text{Cathode (reduction): } Cu^{2+}\,(aq) + 2\,e^- \longrightarrow Cu\,(s)$$

An inherent characteristic of galvanic cells is the external wire that connects the two electrodes, permitting the passage of electrons. This flow of electrons from anode to cathode generates electric current, which can be harnessed to perform work—a principle fundamental to battery technology.

A key component of galvanic cells is the salt bridge, a tube containing an electrolyte solution that completes the circuit by allowing ions to flow between the two half-cells. This migration of ions balances the charge difference that arises as electrons move through the external circuit, preventing charge buildup that would otherwise cease the reaction.

The efficiency and feasibility of galvanic cells rely upon several factors, primarily the cell potential (EMF), calculated based on the reduction potentials of the half-cell reactions. The magnitude of the EMF highlights the potential difference between the electrodes, influencing the voltage and power output of the cell.

Applications of galvanic cells are widespread, notably in the development and utilization of batteries. Primary batteries, such as alkaline batteries, and secondary batteries, like lithium-ion rechargeable batteries, are the mainstay of

9.3. GALVANIC AND ELECTROLYTIC CELLS

portable power solutions globally. Fuel cells, another variant of galvanic cells, convert chemical energy from fuels like hydrogen into electricity through electrochemical reactions, offering efficient and clean energy alternatives.

In contrast, electrolytic cells function by using electrical energy to drive non-spontaneous chemical reactions. This process is the reverse of galvanic cells, involving the external application of voltage to induce chemical changes that would not occur naturally. Electrolysis, the process carried out by electrolytic cells, is vital for various industrial and chemical processes.

A quintessential example of an electrolytic cell is the decomposition of water into hydrogen and oxygen gases. This is accomplished by applying an external voltage across electrodes submerged in water, typically with an added electrolyte to enhance conductivity. The half-reactions are:

$$\text{Anode (oxidation): } 2\,H_2O\,(l) \longrightarrow O_2\,(g) + 4\,H^+\,(aq) + 4\,e^-$$

$$\text{Cathode (reduction): } 4\,H^+\,(aq) + 4\,e^- \longrightarrow 2\,H_2\,(g)$$

In this electrolytic arrangement, water is effectively split into its constituent gases, with oxygen evolving at the anode and hydrogen at the cathode. The energy required for this reaction typically exceeds the energy output from the recombination of hydrogen and oxygen, highlighting the non-spontaneous nature of electrolytic processes.

The practical use of electrolytic cells extends to electroplating, a process where a metal layer is deposited onto a conductive surface. By immersing an object in a solution containing metal ions and applying an electric current, the object acts as the cathode, receiving a coating of the desired metal over time. This technique is utilized extensively in manufacturing, jewelry crafting, and as a corrosion-resistant measure.

Electrolytic cells are also pivotal in the extraction and refining of metals, such as electrorefining of copper, where impure copper is purified through controlled electrolysis. The anode comprises impure copper, while pure copper is deposited at the cathode from the circulating copper ions, leading to high-purity metal production vital for electronics and conductivity applications.

A comprehensive understanding of galvanic and electrolytic cells involves examining their thermodynamic and kinetic aspects. In galvanic cells, the spontaneity of reactions is characterized by negative Gibbs free energy ($\Delta G < 0$), indicating a favorable process that can do work on its surroundings. Conversely, the non-spontaneity in electrolytic cells is marked by positive Gibbs free energy ($\Delta G > 0$), requiring external work to overcome natural tendencies.

The efficiency of both cell types can be influenced by variables such as temperature, concentration, and pressure, factors that modify the chemical equilibrium and reaction rates. Optimization of these parameters forms the basis of designing cells for specific applications, balancing factors like energy efficiency, cost, and environmental impact.

Research continues to enhance the performance and sustainability of electrochemical cells. Innovations in materials science, especially with novel electrode materials and improved electrolytes, have significantly impacted the cell life, energy density, and charge-discharge cycles of modern batteries. Similarly, advancements in electrolytic processes aim to lower energy consumption and minimize environmental footprints, notably in hydrogen production for fuel cells and carbon capture technologies.

Galvanic and electrolytic cells serve as the dual pillars upon which much of modern electrochemistry rests. Their diverse applications, from powering consumer electronics to industrial synthesis of chemicals, underscore a deep-seated dependence on the efficient management of reactions that interconvert energy forms. The ever-expanding role of these cells in sustainable technologies will likely continue, driven by ongoing research and development to meet the energy challenges of the future.

9.4 Conductivity and its Measurement

Conductivity is a critical property of materials that gauges their ability to conduct electric current. In electrochemistry, the focus often lies on ionic conductivity, which involves the movement of ions through an electrolyte solution. Understanding and accurately measuring conductivity are essential for a multitude of scientific, industrial, and environmental applications, including quality control in water systems and the development of sensors.

The concept of conductivity (κ) is quantitatively expressed as the reciprocal of resistivity (ρ) and measured in units of siemens per meter (S/m). Conductivity reflects the ease with which ions move under the influence of an electric field, directly correlating to the concentration and mobility of ions in the solution.

9.4. CONDUCTIVITY AND ITS MEASUREMENT

$$\kappa = \frac{1}{\rho}$$

In electrolytic solutions, conductivity depends on the type and concentration of ions, the temperature, the viscosity of the solvent, and the interaction between the ions and the solvent molecules. Consider a basic salt solution, such as sodium chloride (NaCl). When dissolved in water, NaCl dissociates into Na^+ and Cl^- ions, which migrate in response to an applied electric field, thus contributing to the solution's overall conductivity.

Factors affecting ionic conductivity in solutions include:

- **Ion Concentration:** Conductivity generally increases with greater concentration of ions. However, at high concentrations, interactions between ions may lead to ion pairing or clustering, which can reduce the net mobility and, consequently, the conductivity.

- **Molar Conductivity:** Molar conductivity (λ_m) is defined as the conductivity of the solution divided by the molarity of the solute, providing insight into the efficiency of ions in conducting electricity at a given concentration. With increasing dilution, molar conductivity typically rises due to decreased ion interactions and enhanced mobility.

- **Temperature:** As temperature increases, ion mobility improves due to decreased viscosity of the solvent and increased kinetic energy of the ions, generally leading to higher conductivity.

- **Ion Mobility:** This depends on the size, charge, and solvation of ions. Smaller, highly charged ions tend to exhibit higher mobility, although extensive solvation can impede their movement.

- **Nature of Solvent:** Solvent viscosity and polarity can significantly impact ion movement. Polar solvents, like water, facilitate better dissociation and mobility of ions compared to less polar counterparts.

The measurement of conductivity is typically conducted using a conductivity meter, an instrument that calculates conductivity through the application of an alternating current (AC) between two electrodes submerged in the solution:

- **Electrode Configuration:** The electrodes, often platinized platinum to enhance measurement accuracy, are separated by a specific distance, determining the cell constant (K), which is crucial for accurate measurements as it calibrates the instrument to the cell's geometry.

- **Cell Constant** (K): Defined as the ratio of the distance between the electrodes (d) to the cross-sectional area of the electrodes (A), the cell constant allows transformations between observed conductance and traditionally measured conductivity:

$$K = \frac{d}{A}$$

$$\kappa = K \cdot G$$

where G is the conductance (measured in siemens) of the solution between the electrodes.

Conductivity meters employ alternating current to mitigate polarization effects that can distort measurements, ensuring precision across a range of electrolyte concentrations. Advanced meters often adjust for temperature variations using temperature compensation circuits, reporting conductivity values as if measured at a reference temperature (usually $25°C$).

Calibration is a paramount aspect of conductivity measurement, indispensable for ensuring data accuracy. Calibration involves utilizing standard solutions with known conductivity to set baseline measurements for comparison. The reliability of the results leans heavily on stringent adherence to proper calibration protocols.

Conductivity measurements underpin numerous applications, particularly in environmental monitoring, where they serve as rapid indicators of water quality. In environmental science, conductivity measurements help estimate total dissolved solids (TDS) within aqueous systems, offering insights into pollutant levels in natural and treated waters.

$$\text{TDS (mg/L)} \approx \kappa \times \text{factor (unique to each solution type)}$$

Even slight deviations in conductivity can signal significant alterations in water composition, warranting rapid responses to pollution incidents or system failures.

In industrial processes, conductivity monitoring is crucial for maintaining optimal conditions in chemical reactors and managing the quality of food and beverages. For instance, in brewing and agriculture, the salinity inferred from conductivity influences taste and growth parameters, respectively, highlighting conductivity's broad-ranging impact.

Electrochemical sensors leverage precision conductivity readings to analyze ionic strength and the presence of specific ions in complex solutions. These sensors form the basis of various analytical techniques, such as capillary electrophoresis and ion-exchange chromatography, where conductivity detection denotes solute concentrations as ions are separated or exchanged.

As technology progresses, enhanced conductivity measurement techniques continue to emerge. Microfluidic devices, for example, integrate conductivity detection with miniaturized sensor systems enabling high-throughput analysis with reduced sample volumes. Advances in real-time data collection and smart sensing refine conductivity assessments, aligning with the needs of dynamic, modern industries.

The development of novel materials, particularly the utilization of graphene and carbon nanotube-based electrodes, further enhances the sensitivity and range of conductivity meters. These materials, offering lower resistance and higher surface area, improve interaction with ions, broadening applications even within harsh environments where traditional electrode materials degrade rapidly.

Conductivity, a measure of a material's ability to convey electric current, remains a pivotal parameter guiding the comprehensive analysis of electrolyte solutions. From basic laboratory applications to sophisticated environmental assessments and industrial operations, the determination and understanding of conductivity encapsulate a vital component of electrochemical exploration and innovation. The theory and practice of conductivity measurement continue to evolve, adapting responsively to the demands of science and technology while expanding our understanding of ionic interactions within diverse systems.

9.5 Applications of Electrochemistry

Electrochemistry, the branch of science that examines the interplay between electrical and chemical phenomena, has found widespread application across various fields due to its ability to manipulate chemical reactions and energy conversion processes via electron transfer. The applications of electrochemistry are diverse, encompassing energy storage systems, industrial synthesis, analytical techniques, and medical devices, among others. This section explores several key applications, detailing their principles, technological advancements, and future prospects.

One of the most prominent applications of electrochemistry lies in energy storage and conversion, particularly within batteries and fuel cells. Batteries are electrochemical devices that store energy as chemical potential, releasing it as electrical energy through spontaneous redox reactions. The versatility and reliability of batteries have led to their ubiquitous presence in smartphones, laptops, electric vehicles, and grid storage solutions.

Rechargeable batteries, such as lithium-ion batteries, dominate the energy storage landscape due to their high energy densities, long cycle life, and decreasing costs. The electrochemical reactions in a typical lithium-ion cell involve the transfer of lithium ions between the anode and cathode through an electrolyte during charging and discharging cycles:

$$\text{Anode Reaction: } LiC_6 \longrightarrow C_6 + Li^+ + e^-$$
$$\text{Cathode Reaction: } LiCoO_2 + Li^+ + e^- \longrightarrow Li_2CoO_2$$

The efficiency and safety of these batteries rely on ongoing material advancements, such as the development of solid electrolytes to replace flammable liquid electrolytes, and the engineering of high-capacity anode and cathode materials, including silicon and lithium iron phosphate.

Fuel cells, another cornerstone of electrochemical applications, convert chemical energy directly into electrical energy through electrochemical reactions involving fuels such as hydrogen. Unlike batteries, fuel cells produce electricity as long as fuel is supplied, characterized by higher efficiency and lower emissions. The proton-exchange membrane fuel cell (PEMFC) is particularly appealing for transportation and portable power sources:

$$\text{Anode Reaction: } H_2 \longrightarrow 2H^+ + 2e^-$$
$$\text{Cathode Reaction: } O_2 + 4H^+ + 4e^- \longrightarrow 2H_2O$$

Recent developments focus on reducing the reliance on costly platinum catalysts by exploring alternative materials like transition metal compounds and

9.5. APPLICATIONS OF ELECTROCHEMISTRY

carbon-based catalysts to enhance economic viability.

Electrochemistry also plays a pivotal role in industrial synthesis processes, notably in electrolysis and electroplating. Industrial electrolysis, the process of driving non-spontaneous chemical reactions using electricity, is essential in producing key chemicals. Water electrolysis, for example, produces hydrogen and oxygen gases using renewable electricity:

$$2 H_2O \longrightarrow 2 H_2 + O_2$$

This technology supports the production of green hydrogen, a clean energy carrier with the potential to revolutionize energy systems worldwide. Additionally, electrolysis of sodium chloride yields chlorine and sodium hydroxide, vital chemicals in the production of plastics, disinfectants, and paper.

Electroplating, another critical industrial application, involves the deposition of a metal coating onto a substrate through electrochemical reduction. This process is utilized to enhance surface properties such as corrosion resistance, hardness, and aesthetics in jewelry, automotive components, and electronic circuits:

$$M^{n+} + n\,e^- \longrightarrow M\,(s)$$

Advanced techniques employ pulse plating and alloy deposition to refine surface structure and improve quality, reducing waste and environmental impact.

Electrochemical sensors leverage the principles of electrochemistry to detect and quantify chemical species with high sensitivity and specificity. These sensors, encompassing a range of electrodes and potentiometric, amperometric, and conductometric methods, have applications in environmental monitoring, medicine, and food safety.

Enzymatic glucose sensors, a type of amperometric sensor, are prevalent in diabetes management, measuring blood glucose levels through the oxidation of glucose catalyzed by glucose oxidase:

$$\text{Glucose} + O_2 \longrightarrow \text{Gluconolactone} + H_2O_2$$

Amperometric detection of hydrogen peroxide at a working electrode indicates glucose concentration. Advances in sensor technology focus on miniaturization, multisensing capabilities, and integration with digital platforms for enhanced user convenience.

In corrosion protection, electrochemistry offers methods to prevent the deterioration of metallic structures, crucial in infrastructure maintenance and safety. Techniques such as cathodic protection apply a protective current to metal structures, inhibiting oxidation reactions responsible for corrosion:

$$M + O_2 + 2H_2O + 4e^- \longrightarrow 4OH^- + M(s)$$

This is particularly important for pipelines, bridges, and marine vessels exposed to harsh environments. Research into smart coatings and self-healing materials aims to further increase longevity and reduce maintenance costs.

Electrochemical principles extend into the realm of organic synthesis, enabling greener and more efficient routes to valuable compounds. Electrosynthesis involves stoichiometric generation of radicals and transition states under mild conditions, offering precise control over reaction pathways without the need for hazardous reagents. This is of growing interest for the sustainable synthesis of pharmaceuticals, polymers, and fine chemicals.

The versatility of electrochemistry spans numerous applications transforming various sectors, from power generation to advanced materials and biological systems. The continual development of electrochemical technologies aims to address global challenges, including sustainable energy supply, environmental sustainability, and industrial efficiency, positioning electrochemistry at the forefront of scientific and technological innovation. As research progresses, the expansion of electrochemical techniques promises to unlock further possibilities, enriching the interface between chemical processes and electrical energy.

9.6 Electrochemical Techniques and Instrumentation

Electrochemical techniques form the backbone of modern electrochemical analysis, offering insight into a range of chemical processes, from kinetic studies to mechanistic information and quantitative analysis. Advanced instrumentation facilitates these techniques, allowing precise control and measurement of electrochemical reactions in various settings. This section thoroughly

9.6. ELECTROCHEMICAL TECHNIQUES AND INSTRUMENTATION

explores the principal techniques and corresponding instrumentation, emphasizing their applications and significance in scientific research and industrial contexts.

Central to the study of electrochemical reactions is the understanding of the interactions between electrodes and their chemical environment. Techniques like cyclic voltammetry, chronoamperometry, and electrochemical impedance spectroscopy are routinely employed to investigate these interactions, providing critical data on reaction kinetics, mechanisms, and material properties.

Cyclic Voltammetry (CV), a widely used electrochemical technique, involves cycling the potential of a working electrode linearly over time while recording the resulting current. This method is particularly useful for examining redox properties of molecules, determining electrochemical reaction mechanisms, and evaluating electron transfer rates. A typical setup includes a three-electrode system comprising a working electrode, a reference electrode, and a counter electrode.

The working electrode, usually made of materials like glassy carbon, platinum, or gold, is where the redox reaction occurs. The reference electrode, such as a saturated calomel electrode (SCE) or silver/silver chloride electrode (Ag/AgCl), maintains a constant potential, while the counter electrode, often a platinum wire, completes the circuit without limiting current flow.

A cyclic voltammogram provides a wealth of information, with peak currents related to analyte concentration and peak separations offering clues about reaction reversibility. The Randles-Ševčík equation relates peak current (i_p) in a reversible CV to analyte concentration (C), scan rate (v), and other parameters:

$$i_p = (2.69 \times 10^5) n^{3/2} A D^{1/2} v^{1/2} C$$

where n is the number of electrons transferred, A is the electrode area, and D is the diffusion coefficient.

Chronoamperometry (CA), another potent technique, measures the current response of an electrochemical system following a potential step applied to the working electrode. This method is used to study reaction kinetics, especially for systems involving diffusive control. By maintaining a constant potential and recording the resulting transient current, chronoamperometry elucidates details about the diffusion layer and reaction rates.

Fick's laws of diffusion govern the temporal evolution of the current, allowing quantitative analysis of parameters like diffusion coefficients. The Cottrell equation describes the current ($i(t)$) after a potential step under diffusion control:

$$i(t) = \frac{nFAcD^{1/2}}{\sqrt{\pi t}}$$

where F is Faraday's constant, a the initial concentration, and t is time.

Electrochemical Impedance Spectroscopy (EIS) is an advanced technique that analyzes the frequency response of an electrochemical system to a small AC voltage. This method provides comprehensive insights into electron transfer dynamics, reaction rates, and cell impedance, proving invaluable in characterizing materials for batteries, sensors, and coatings.

An EIS experiment involves applying a sinusoidal perturbation to the system while measuring the resulting current across a spectrum of frequencies. The impedance (Z) can be modeled using equivalent electrical circuits composed of resistors, capacitors, inductors, and constant phase elements to delineate processes like charge transfer, double-layer capacitance, and diffusion.

The Nyquist plot, a common graphical representation of EIS data, displays impedance as imaginary versus real components. The analysis of these plots aids in extrapolating physical parameters, offering a non-invasive view of internal processes without destroying samples.

Instrumentational advances have enhanced the accuracy and resolution of electrochemical measurements. Potentiostats, essential devices that maintain desired potentials between electrodes, are integral to conducting electrochemical experiments. Modern potentiostats incorporate digital interfaces, software-controlled experiments, and data processing capabilities, allowing complex experiments and rapid parameter extraction. Additionally, multiplexed or multichannel systems enable simultaneous investigation of multiple samples, facilitating high-throughput screening and parallel analysis.

Electrochemical sensors and biosensors, leveraging the aforementioned techniques, have gained prominence due to their high sensitivity and specificity. By converting biochemical interactions into electronic signals, these sensors facilitate real-time monitoring of analytes in medical diagnostics, environmental analysis, and food safety.

9.6. ELECTROCHEMICAL TECHNIQUES AND INSTRUMENTATION

Biosensors, specifically, integrate a biological recognition element (enzyme, antibody, or nucleic acid) with a transducer, producing a quantifiable electrochemical response. Glucose sensors, for example, exploit the enzyme glucose oxidase to oxidize glucose, with the resultant product generating a measurable current proportional to glucose concentration.

In environmental applications, electrochemical sensors detect heavy metals, nitrogen compounds, and organic pollutants with precision. Enhancements in sensor technology, such as immobilizing enzymes onto nanostructured electrodes or utilizing aptamer-based recognition, continually expand their practical utility and analytical performance.

Laboratory-on-chip (LOC) devices represent another frontier in electrochemical instrumentation, miniaturizing and integrating multiple laboratory functions onto a single chip. These microfluidic platforms facilitate small-volume, multiplexed analyses with rapid turnaround times, paving the way for point-of-care diagnostics and personalized medicine.

Fundamental to translating laboratory advances into real-world applications is the development of materials and interfaces that improve electrocatalysis and electroanalytical performance. Nanostructuring of electrode surfaces, through methods such as electrodeposition and self-assembly, enhances electron transfer rates and reactive surface area, yielding faster response times and lower detection limits.

Research continues to advance the potential of electrochemical techniques across vast application areas. Emerging fields like bioelectrochemistry explore the interface between electronics and living systems, focusing on energy harvesting from biological processes and developing biohybrid devices.

Electrochemical techniques and instrumentation stand as linchpins in modern analytical and physical chemistry. Their ability to decode complex interactions at interfaces aids in uncovering fundamental reaction dynamics and advancing technological development across various fields, from energy conversion to biomedical engineering. As techniques evolve, the breadth and depth of electrochemical applications are set to expand further, supporting the ever-changing demands of science and industry.

9.7 Ionic Conductivity and Transport Phenomena

Ionic conductivity and transport phenomena are essential concepts in electrochemistry that describe the movement of ions within electrolytic solutions and solid-state materials. This movement underlies the fundamental operations of diverse electrochemical systems from batteries and fuel cells to biological membranes and industrial processes. In this section, we explore the intricacies of ionic conductivity, the mechanisms of ionic transport, and the variables affecting these phenomena.

Ionic conductivity (σ) measures how effectively ions carry electric charge through a medium. It is defined as:

$$\sigma = nq\mu$$

where n is the number density of mobile charge carriers (ions), q is the charge of the ions, and μ is their mobility. High ionic conductivity typically implies a high number of charge carriers with significant mobility, generally found in environments with dissociated salts as well as in some solid electrolytes.

Ionic transport occurs via diffusion and migration. *Diffusion* is the movement of ions from regions of high concentration to low concentration, governed by Fick's laws. On the other hand, *migration* involves ion movement in response to an electric field. Both mechanisms often operate simultaneously, especially in electrochemical cells, where the exchange and balance of charge are essential for sustained operation.

- Diffusion in Electrolytic Solutions

Fick's first law describes steady-state diffusion, where the flux of ions (J) is proportional to the concentration gradient:

$$J = -D\frac{\partial C}{\partial x}$$

where D is the diffusion coefficient, C the concentration, and x the spatial coordinate. The proportionality indicates how effectively ions spread through a medium.

9.7. IONIC CONDUCTIVITY AND TRANSPORT PHENOMENA

Fick's second law addresses non-steady state diffusion, predicting how concentration changes with time:

$$\frac{\partial C}{\partial t} = D\frac{\partial^2 C}{\partial x^2}$$

These equations serve as the basis for analyzing concentration profiles within electrolytic and ionic systems, proving particularly useful in characterizing the time-dependent response of electrochemical devices.

In electrochemical cells, diffusion layers form adjacent to electrodes. Here, concentration gradients develop as reactive species are consumed or produced, influencing device performance. Analytical techniques, including chronoamperometry, exploit these gradients to extract information about reaction dynamics and transport properties.

- Migration in Electric Fields

In response to an applied electric field, charged species migrate, experiencing forces proportional to their charge and the field strength. The migration velocity (v) of ions is given by:

$$v = \mu E$$

where E is the electric field strength. Ionic mobility (μ), dependent on the ion size, charge, and interaction with the solvent, dictates how swiftly ions respond to the field.

The Nernst-Planck equation consolidates both diffusion and migration effects, providing a comprehensive model of ion transport in electrochemical systems:

$$J_i = -D_i\frac{\partial C_i}{\partial x} + z_i\mu_i C_i E$$

Here, J_i is the flux of ion species i, D_i is its diffusion coefficient, z_i its charge number, and μ_i its mobility.

- Transport in Solid Electrolytes

Ionic transport in solid-state electrolytes opens avenues for all-solid-state batteries and fuel cells. Unlike liquid electrolytes, solid-state electrolytes offer advantages like leakage prevention, flammability reduction, and thermal stability. However, achieving high ionic conductivity in these materials poses a challenge due to their typically rigid lattice structures.

In solid electrolytes, ionic movement occurs via vacancies or interstitial sites in crystalline networks. Materials such as garnet-type lithium lanthanum zirconates for lithium-ion conduction, and NASICON-type materials for sodium ions, exemplify advances in this field. Efforts continue to synthesize novel materials with enhanced ionic conductivity by controlling defect concentrations and lattice symmetries, supporting next-generation energy storage technologies.

- Factors Affecting Ionic Conductivity

Various factors influence ionic conductivity:

- **Concentration of Ions**: Good conductivity generally requires sufficient ion availability. However, ion-ion interactions at high concentrations can hinder movement through clustering or pairing, reducing effective conductivity.

- **Temperature**: Conductivity usually rises with temperature due to increased ion kinetic energy and lower viscosity, promoting easier mobility.

- **Solvent Properties**: Solvent viscosity and polarity affect ion solvation and, consequently, ion mobility. In solid electrolytes, lattice dynamics, particle size, and grain boundaries are crucial.

- **Ion Valency and Size**: Highly charged ions move less freely due to strong electrostatic attractions, while larger ions experience higher steric hindrances through constrained pathways.

- **The Presence of Complexes**: Complex formation with counter-ions or solvent molecules can trap ions, reducing their mobility.

- Applicative Aspects

Batteries and Fuel Cells: The optimization of ionic conductivity is central to the advancement of rechargeable battery technologies. Enhancing ion transport within electrolyte materials boosts energy density and battery efficiency. For fuel cells, proton-conducting membranes, like Nafion in PEM fuel cells, are integral, requiring fine-tuning to ensure performance, especially under varying humidity and temperature conditions.

Sensors and Actuators: Ionic conductors facilitate the actuation and response in electrochemical sensors, where precise ion exchange drives signal transduction. Development of sensors with responsive ionic conductors enables fast detection and manipulation of ionic species across biological and environmental media.

Ionic Liquids and Polyelectrolytes: These unique materials, comprising densely ionic functional groups, provide versatile platforms for electrochemical applications, from electrolytes to actuators. Tailoring their viscoelastic and conductivity properties further extends their utilization in technological applications.

Biological Systems: Within biological systems, ionic conductivity is critical for nerve impulse propagation. The opening and closing of ion channels selectively allow ions like sodium, potassium, and calcium to flow across membranes, creating action potentials necessary for neuronal communication.

- Future Directions

The ongoing exploration of ion transport phenomena and conductivity in both traditional and novel systems holds promise for impactful innovations. Research into hybrid electrolytes, combining the advantages of liquid and solid systems, aims to harness dual benefits of flexibility and stability. Furthermore, scalable manufacturing and integration of high-conductivity materials into existing infrastructure remain key for technological breakthroughs.

The convergence of computational modeling, materials science, and electrochemical engineering enables precise control over ionic transport properties, ultimately fostering developments in sustainability-driven technologies, medical treatments, and smart devices. Ionic conductivity and transport phenomena form an integral link in transforming fundamental electrochemical science into actionable technological innovations, establishing pathways to address diverse and challenging global needs.

Chapter 10

Materials Chemistry and Nanotechnology

Materials chemistry and nanotechnology intersect at the forefront of scientific innovation, exploring the synthesis, characterization, and application of materials with dimensions at the nanoscale. This chapter delves into the diverse properties of metals, ceramics, polymers, and composites, emphasizing the structure-property relationships that dictate their macroscopic behaviors. Through cutting-edge techniques for synthesizing nanomaterials, the chapter highlights advancements in manipulating matter at the atomic level to engineer materials with tailored characteristics. Characterization methods such as electron microscopy and spectroscopy provide intricate details about these materials, enabling precise control over their functionalities. The exploration extends to real-world applications in electronics, medicine, and environmental technology, demonstrating the transformative potential of nanotechnology in addressing complex global challenges.

10.1 Principles of Materials Chemistry

The field of materials chemistry encompasses the study of the chemical processes and methodologies required to create new materials with unique and tailored properties. It is an interdisciplinary science that draws from chemical engineering, physics, and materials science to explore the synthesis, characterization, and application of chemical compounds. The principles of materials chemistry are founded upon understanding the composition of materials, analyzing their intrinsic properties, and devising methodologies for their fabrication and modification.

Central to materials chemistry is the exploration of the structure of materials on atomic and molecular scales. The arrangement of atoms and molecules dictates the characteristics such as mechanical strength, thermal stability, electrical conductivity, and optical properties. Materials chemists employ various characterization techniques to determine structural parameters, including crystallinity, phase composition, and molecular symmetry.

A core concept within materials chemistry is the relationship between the structure and properties of a material. The structure-property paradigm posits that the microscopic structure of a material informs its macroscopic functionalities. For instance, the tightly packed lattice arrangement in metals like iron ensures excellent electrical conductivity and malleability, whereas the covalent network structure of a ceramic such as silicon carbide results in exceptional hardness and thermal resistance. Understanding this interrelationship enables scientists to engineer materials with specific attributes by manipulating their underlying structure through chemical processes.

Another fundamental principle involves the thermodynamics and kinetics of material formation. Thermodynamics provides insight into the stability of materials and influences the direction and extent of chemical reactions. Kinetics, on the other hand, governs the speed at which these reactions proceed, affecting the final microstructure. Thus, an adept materials chemist must balance thermodynamic favorability with kinetic constraints to optimize the synthesis of materials. Techniques such as solvothermal synthesis, vapor deposition, and sol-gel processing are employed to achieve precise control over reaction environments, guiding the development and stabilization of new materials.

Materials chemistry also explores the properties and dynamics of interfaces and surfaces, which play crucial roles in determining a material's performance in a given application. Interfaces are particularly significant in composite materials where different phases interact, and at the nanoscale where surface-to-volume ratios become substantial. Chemical modifications of surfaces, such as functionalization with organic ligands or with oxides, can dramatically enhance properties like corrosion resistance or catalytic activity.

10.1. PRINCIPLES OF MATERIALS CHEMISTRY

The design and discovery of new materials are also grounded in computational materials science. Quantum mechanical calculations, molecular dynamics simulations, and computational thermodynamics are increasingly valuable tools that provide predictive insights into material behavior before experimental realization. Such approaches expedite the discovery process, allowing for computational screening of potential materials to identify those with the most promising properties for specific applications.

Moreover, materials chemistry is a pivotal field in the development of sustainable technologies. Green chemistry principles are increasingly integrated into materials design to minimize environmental impact and enhance recyclability. This involves the use of renewable resources, energy-efficient synthesis processes, and the development of non-toxic materials. Innovations such as biodegradable polymers and non-persistent organic pollutants are a testament to the impactful convergence of sustainability and materials chemistry.

Materials characterized by their electronic, magnetic, and optical properties are of particular interest. Semiconductors, which are foundational to modern electronic devices, are a primary focus. The ability to control electronic band structures through doping and epitaxial growth leads to tunable electronic and optical properties, essential for the functionality of transistors, solar cells, and light-emitting diodes.

In addition to their electronic applications, materials chemistry contributes significantly to the energy sector. The design of advanced materials for energy storage and conversion, such as lithium-ion batteries and photovoltaic cells, relies on understanding the electrochemical properties of materials. For example, the choice of electrode materials and the electrolyte composition in battery technology affect both the energy density and cycle life, translating directly to performance and durability.

The principles of materials chemistry are not static; they continuously evolve with advances in theory, experimentation, and technology. This evolution necessitates a commitment to lifelong learning and adaptation within the field, ensuring that new paradigms are rapidly integrated into existing frameworks for material development.

Nanoscale materials offer a revolutionary perspective on traditional materials chemistry, introducing unique phenomena absent at larger scales due to quantum mechanical effects and surface energy considerations. Nanomaterials, such as nanoparticles, nanowires, and quantum dots, exhibit size-dependent properties that diverge considerably from their bulk counterparts. These properties open avenues for innovations in catalysis, where enhanced surface areas and active sites increase reaction efficiencies, and in medicine, where targeted drug delivery systems can be devised.

Chemical bonding and interactions are also crucial elements in the study of materials. The character of the bonds—ionic, covalent, metallic, or van der Waals—affects everything from melting point to electrical conductivity. Advanced spectroscopic techniques, such as nuclear magnetic resonance (NMR) and X-ray photoelectron spectroscopy (XPS), are routinely employed to study these interactions and provide insights into the bonding environment within a material.

In practice, the field of materials chemistry interacts closely with engineering disciplines to translate scientific discovery into commercializable technology. This involves scaling up laboratory synthesis routes into industrial processes and ensuring the reliability and viability of materials used in manufacturing. Pioneers in materials chemistry must, therefore, collaborate across disciplines to address the economic, technical, and societal challenges presented by new materials.

Finally, as the field progresses, ethical considerations and the societal implications of materials chemistry garner increasing attention. Issues such as resource scarcity, pollution, and the potential toxicity of new materials must be addressed. Comprehensive risk assessment and life-cycle analysis are essential tools to anticipate and mitigate the negative impacts of new material technologies.

The principles of materials chemistry constitute a richly interconnected tapestry of scientific knowledge, combining historical techniques with state-of-the-art advancements to address contemporary challenges. The discipline holds vast potential in revolutionizing industries and contributing to the sustainable development of society. Through continual interdisciplinary collaboration and integration of innovative methodologies, materials chemistry will remain pivotal in shaping the technological landscape of the future.

10.2 Types of Materials: Metals, Ceramics, Polymers, and Composites

Materials science classifies materials into broad categories based on their chemical composition, structural characteristics, and intrinsic properties. Among the primary classes of materials are metals, ceramics, polymers, and composites. Each classification is distinguished by unique attributes that lend themselves to specific applications, taking into account factors such as mechanical strength, thermal and electrical conductivity, ductility, and resistance to corrosion or degradation.

Metals and their alloys are foundational to modern industrial applications due to their malleability, ductility, and excellent electrical and thermal conductivity. The metallic bond, characterized by a sea of delocalized electrons, is responsible for many of the desirable properties of metals. Within the category of metals, pure metals such as copper, aluminum, and iron are often utilized for specific applications—copper for its superior electrical conductivity, aluminum for its low density and strength-to-weight ratio, and iron primarily in the form of steel for its tensile strength and versatility.

Metal alloys, created by combining two or more metallic elements or metals with non-metals, acknowledge the limitations of pure metals and offer improved mechanical and physical properties. Alloys such as stainless steel, brass, and bronze exhibit enhanced corrosion resistance, strength, and aesthetic appeal compared to their constituent metals alone. The manipulation of alloy composition and heat-treatment processes allows metallurgists to tailor these materials for applications ranging from construction to aerospace.

Ceramics are inorganic, non-metallic materials typically formed by a process of heating and subsequent cooling. They exhibit exceptional hardness, high melting temperatures, and chemical stability. The ionic and covalent bonding in ceramics creates strong directional bonds, providing them with durability and resistance to mechanical and thermal stress. Prominent examples of ceramics include silicon carbide, alumina, and zirconia, each playing integral roles in industries ranging from electronics to construction to medical devices due to their desirable insulating properties and resistance to high temperatures.

However, ceramics have inherent brittleness due to their rigid atomic lattice structures, leading to catastrophic failure under tensile stress. Innovative processing techniques, such as toughening methods that introduce compressive surface stresses, have been developed to mitigate this vulnerability, resulting in materials like yttria-stabilized zirconia, known for its enhanced fracture toughness.

Polymers, which encompass a vast range of synthetic and natural substances composed of large, repeating molecular chains, are another critical class of materials. Their characteristic properties, including low density, high flexibility, and resistance to corrosion, arise from the long chains of covalently bonded atoms that move freely past one another due to weak Van der Waals forces.

Synthetic polymers like polyethylene, polystyrene, and polyvinyl chloride (PVC) have transformed everyday life through their application in packaging, textiles, and construction. The versatility of polymers is showcased in their ability to be modified and adapted, resulting in specific properties and a wide variety of commercial uses. For example, variations in polymerization methods and copolymer formation can create materials with tailored mechanical properties such as thermoplastics and thermosetting plastics.

A significant aspect of polymer science is the study of polymer composites, which involve reinforcing a polymer matrix with fibers or particulate fillers to enhance mechanical performance and thermal stability. Composite materials capitalize on the strengths of their constituents to form a superior material capable of addressing the limitations of any single component, making them an increasingly attractive choice in industries such as automotive, aerospace, and sports equipment.

Composites, as diverse and complex assemblies of materials, are formed by combining two or more constituent materials with notably different physical or chemical properties to produce a material with characteristics different from each individual component. The components of a composite typically include a continuous phase (matrix) and a dispersed phase (reinforcement). The matrix phase may comprise metals, polymers, or ceramics, while the reinforcement material can be in the form of fibers, whiskers, or particulates.

One prominent example of a composite is fiberglass, composed of a polymer matrix reinforced with glass fibers, offering high strength and lightweight properties, ideal for automotive and marine industries. Carbon fiber-reinforced polymers (CFRPs), where carbon fiber serves as the reinforcement within a resin matrix, are celebrated for their exceptional stiffness-to-weight ratios, making them indispensable in high-performance applications such as aerospace engineering and competitive sports equipment.

10.3. STRUCTURE-PROPERTY RELATIONSHIPS

Additionally, hybrid composites that incorporate multiple reinforcing materials within a single matrix system provide further opportunities for material customization, enabling specific combinations of mechanical, thermal, or electrical properties not possible with traditional composites. The development of such advanced materials is driven by a growing demand for eco-friendliness, recyclability, and cost-effectiveness without compromising performance.

The intrinsic properties of each class of materials allow engineers and scientists to exploit their characteristics and optimize their use in a broad range of engineering designs and processes. Multiscale modeling and simulation play an increasingly significant role in understanding and predicting the behavior of complex materials like polymers and composites, offering insights into their deformation, failure mechanisms, and performance under various conditions.

As materials science continues to advance, the intersection of the fundamental principles underlying metals, ceramics, polymers, and composites with emerging technologies highlights the need for continual research and development. Challenges such as material fatigue, environmental impact, and resource scarcity drive innovation towards creating next-generation materials that are lighter, stronger, and more sustainable, meeting the demands of a rapidly evolving world.

10.3 Structure-Property Relationships

The concept of structure-property relationships lies at the heart of materials science, forming a critical underpinning that enables the understanding and prediction of material performance based on their microstructural characteristics. This concept is rooted in the idea that the properties of materials, whether mechanical, electrical, thermal, or optical, are fundamentally governed by their internal structures, from atomic and molecular scales to macroscopic levels. By dissecting the intricate connections between structure and properties, materials scientists and engineers can systematically design and optimize materials for complex applications.

At the atomic scale, the arrangement and type of atoms determine fundamental properties such as bonding characteristics and chemical reactivity. Materials can exhibit vastly different properties based on small variations such as crystal lattice arrangement, presence of defects, or chemical composition. For instance, carbon atoms can form diamond, with a tetrahedral crystal structure that results in extreme hardness and high thermal conductivity, or graphite, with a planar hexagonal lattice providing lubricity and excellent electrical conductivity.

Crystal structures play a pivotal role in defining the physical attributes of materials. The geometry and symmetry of a crystal lattice influence anisotropic properties—meaning properties that differ along different directions. Anisotropy is evident in materials like single-crystal silicon, crucial for microelectronics, where precision control over electronic properties is achieved by exploiting crystallographic orientation. Advanced characterization techniques such as X-ray diffraction (XRD) allow scientists to probe these structural details, enabling the tailoring of materials through controlled synthesis and processing.

The presence of defects, including point defects such as vacancies and interstitials, dislocations, and grain boundaries, profoundly affects material properties. While defects are often considered imperfections, strategically engineered defects can enhance material functionality. For example, doping semiconductors with impurity atoms introduces electronic states that act as carriers, critically influencing electrical conductivity and forming the basis for modern semiconductor technology. Similarly, dislocations, as linear defects within a crystal structure, play a critical role in dictating the mechanical behavior of metals by serving as pathways for plastic deformation.

Grain boundaries, the interfaces between crystallites in polycrystalline materials, significantly influence mechanical and thermal properties. The size and orientation of these grains can be manipulated through processes such as annealing to enhance toughness and reduce creep in metals. In ceramics, controlling grain structure can optimize properties like thermal shock resistance and dielectric constant.

At the mesoscale, the microstructure of a material reveals additional complexity through the arrangement of multiple phases, secondary phase inclusions, and porosity. Composite materials often rely on engineered microstructures to combine properties disparate in individual constituents, like high strength and low density, achieved in fiber-reinforced polymers (FRP). Techniques such as scanning electron microscopy (SEM) provide insights into microstructural features, allowing for direct observation and manipulation of material characteristics.

Polymers exhibit unique structure-property relationships due to their versatile and complex macromolecular networks. The chain configuration and tacticity—referring to the relative stereochemistry of adjacent chiral centers—

10.3. STRUCTURE-PROPERTY RELATIONSHIPS

affect crystallinity, mechanical flexibility, and thermal behavior. Crystallinity in polymers, which arises from the orderly packing of molecular chains, markedly influences mechanical properties such as tensile strength and elasticity. Techniques like differential scanning calorimetry (DSC) and X-ray scattering elucidate these structural features, guiding the synthesis of commercially applicable materials.

In nanomaterials, where dimensions fall below 100 nanometers, quantum confinement effects introduce novel properties not observed in bulk materials. For instance, quantum dots exhibit size-dependent optical emission properties, enabling their use in applications like biological imaging and display technologies. The large surface area-to-volume ratio in nanoparticles enhances chemical reactivity and catalytic efficiency, forming the basis for innovations in catalysis and medical therapies.

Materials with tunable properties, such as shape memory alloys and piezoelectric materials, further illustrate the power of the structure-property paradigm. Shape memory alloys, like nitinol, exhibit a phase transformation that allows them to return to a predefined shape upon heating. This reversible transformation is a direct consequence of the material's microstructural configuration and makes them invaluable in applications ranging from biomedical devices to actuators.

Moreover, piezoelectric materials, which convert mechanical stress to electrical charge, rely on non-centrosymmetric crystal structures. Their electromechanical coupling is harnessed in sensors, actuators, and energy-harvesting devices. The atomic displacement within the lattice under external stress is key to their operation, emphasizing the fundamental connection between symmetry, structure, and function.

At the macroscopic level, the design and processing of materials incorporate these foundational insights to achieve desired properties in finished products. High-performance steels, for instance, undergo specific thermal and mechanical treatments—quenching, tempering, and cold working—to refine microstructure and achieve optimal strength and ductility.

The emergence of computational materials science marks a new era in exploring and designing materials based on structure-property relationships. First-principles calculations, molecular dynamics simulations, and multiscale modeling provide powerful tools for predicting material behavior and guiding experimental design before synthesis. These computational approaches accelerate the discovery of novel materials with tailored properties for specialized applications, such as high-temperature superconductors and metamaterials designed with negative refractive indices.

Metamaterials, engineered to have properties not found in naturally occurring materials, underscore the triumph of the structure-property relationship paradigm. By designing artificial structures with subwavelength periodicity, scientists create materials with unprecedented optical, acoustic, and electromagnetic phenomena, including cloaking and perfect lensing.

Additionally, the integration of machine learning and neural networks into materials science sets the stage for autonomous material design. By leveraging vast datasets, machine learning algorithms can identify patterns and correlations across different scales, predicting structure-property relationships that might escape traditional empirical methods.

The exploration of structure-property relationships provides a robust framework for understanding the multifaceted behavior of materials. This systematic approach empowers the rational design and optimization of materials across numerous fields, from nanotechnology to structural engineering, by connecting the dots between microscopic structure and macroscopic performance. Through ongoing research and technological advancements, the profound potential of structure-property relationships will continue to drive innovative solutions, addressing the complex challenges of the modern world.

10.4 Synthesis and Fabrication of Nanomaterials

The synthesis and fabrication of nanomaterials encompass a multitude of techniques and approaches aimed at creating materials with unique structures and properties arising at the nanoscale. Given the increasing demand for these materials in various technological and industrial sectors, understanding the fundamental principles and strategies for synthesizing nanomaterials is paramount. These techniques can be broadly categorized into two main approaches: top-down and bottom-up, each associated with distinct methodologies and resulting structures.

The top-down approach involves the reduction of bulk materials to the nanoscale by employing techniques such as mechanical milling, lithography, and ion beam processing. Mechanical milling, a quintessential top-down method, entails physically breaking down bulk materials into nanoparticles through high-energy ball milling. This technique is advantageous due to its

10.4. SYNTHESIS AND FABRICATION OF NANOMATERIALS

simplicity and cost-effectiveness, but it can introduce undesirable defects and a wide particle size distribution.

Advanced lithographic techniques, including photolithography, electron beam lithography, and nanoimprint lithography, provide a more controlled fabrication process. Photolithography utilizes light to transfer geometric patterns onto a substrate, forming the foundation of modern semiconductor device fabrication. Advances in lithography, such as extreme ultraviolet (EUV) lithography, enable the production of ever-smaller features, essential for integrated circuit miniaturization.

Electron beam lithography bypasses diffraction limits faced in photolithography by using a focused beam of electrons to directly write patterns with nanometer precision. While slower and more expensive than photolithography, electron beam lithography is invaluable for prototyping and producing high-resolution structures needed in research and specialized applications.

In contrast, the bottom-up approach focuses on assembling nanomaterials atom-by-atom or molecule-by-molecule from smaller units. This method facilitates the creation of materials with high precision and tailored functionalities. Chemical vapor deposition (CVD) and molecular beam epitaxy (MBE) are hallmark techniques in this category.

Chemical vapor deposition involves the decomposition or chemical reaction of vapor-phase precursors on a heated substrate, allowing for the growth of high-purity thin films and nanostructures. CVD is instrumental in producing carbon nanotubes, silicon nanowires, and other advanced nanostructures with controlled composition and morphology. Variants such as plasma-enhanced CVD introduce plasma to enhance reaction rates and allow for lower processing temperatures, broadening the range of achievable materials.

Molecular beam epitaxy offers unmatched control over layer composition and thickness, crucial for fabricating semiconductor heterostructures and quantum wells. In MBE, atomic or molecular beams of source materials are directed onto a substrate under ultra-high vacuum conditions, allowing for layer-by-layer buildup. The precision of MBE facilitates the exploration of quantum mechanical phenomena and the development of optoelectronic devices.

Solution-based synthesis methods are also prominent within the bottom-up approach, providing versatility and scalability. Sol-gel processing, a popular technique, involves the transition of a solution of organometallic precursors into a networked gel through hydrolysis and polycondensation reactions. The subsequent drying and thermal treatment yield inorganic networks or oxide nanoparticles with high surface areas, used in catalysis, optics, and electronics.

Colloidal synthesis allows for the controlled creation of nanoparticles and nanocrystals by tailoring reaction conditions such as temperature, concentration, and surfactant presence. These procedures enable the production of quantum dots, renowned for their size-dependent optical properties, utilized extensively in display technologies and biological imaging.

Another noteworthy method within the bottom-up family is the self-assembly of molecules or small particles into ordered nanostructures driven by non-covalent interactions, such as hydrogen bonding, van der Waals forces, or ionic interactions. Self-assembly processes facilitate the creation of complex nanostructures and have pivotal applications in developing metamaterials, drug delivery systems, and nanolithography.

Biological synthesis, which leverages natural biological processes to produce nanomaterials, is gaining traction due to its eco-friendliness and biocompatibility. Organisms such as bacteria, fungi, and plants have intrinsic abilities to reduce metal ions to form nanoparticles. This bioinspired approach opens avenues for sustainable synthesis and the expansion of nanotechnology into medical and environmental fields.

The fabrication of nanomaterials is inherently linked to the challenges of achieving uniformity, repeatability, and scalability. Controlling size, shape, and surface characteristics is crucial to tailoring the functionality and enhancing the performance of nanomaterials in practical applications. Precise control over these parameters is achieved through multi-step procedures involving stabilization agents, capping molecules, and kinetic control strategies.

Characterization techniques play an equally critical role in understanding and improving synthesis and fabrication processes. Advanced microscopy tools, such as transmission electron microscopy (TEM) and atomic force microscopy (AFM), provide detailed insights into the morphology and crystallographic orientation of nanoparticles. In tandem, spectroscopy techniques like Raman spectroscopy and X-ray photoelectron spectroscopy (XPS) yield information regarding composition, bonding states, and surface chemistry.

To integrate these nanomaterials into devices and systems, deposition and patterning techniques such as spin coating, dip coating, and inkjet printing are employed. These methodologies facilitate the transition from individual nanoparticles or films to functional components within electronic, sensor, and photovoltaic devices.

Emerging techniques, like atomic layer deposition and additive manufacturing, are further revolutionizing the field. Atomic layer deposition allows for the conformal coating of materials with atomic layer precision, essential for creating complex three-dimensional architectures. Additive manufacturing or 3D printing, extended to the nanoscale, showcases the integration of nanomaterials into intricate, customized geometries with applications ranging from electronics to biomedical implants.

The synergy of synthesis techniques and fabrication strategies marks the continuous evolution of nanotechnology, pushing the boundaries of what is possible in the design and deployment of nanomaterials. Collaboration across disciplines—chemistry, physics, materials science, and engineering—accelerates innovation in this dynamic field, addressing global challenges through advancements in energy storage, environmental remediation, and health care solutions.

As the synthesis and fabrication of nanomaterials mature, ethical considerations, environmental impact, and regulatory frameworks must also evolve. Nanomaterials' small sizes and high reactivity raise questions about toxicity, persistence in the environment, and implications for human health. Researchers and policymakers engage in active discourse to ensure responsible development aligned with societal values and sustainable practices.

The synthesis and fabrication of nanomaterials is a multifaceted and rapidly progressing domain that provides the foundational tools for exploiting the unique properties of materials at the nanoscale. By marrying traditional techniques with groundbreaking innovations, it opens pathways for transforming concepts into reality, elevating technologies to new heights, and addressing the intricacies of modern scientific and engineering challenges.

10.5 Characterization Techniques for Nanomaterials

The characterization of nanomaterials encompasses a diverse array of techniques aimed at probing the physical, chemical, and structural attributes of materials at the nanoscale. Understanding these properties is crucial for optimizing synthesis conditions, elucidating structure-property relationships, and ensuring reliable performance in applications spanning electronics, medicine, and energy. Comprehensive characterization entails measuring particle size, morphology, surface chemistry, crystallographic structure, optical properties, and more. This section delves into the principal techniques leveraged by scientists to unravel the intricate details of nanomaterials.

One of the fundamental tools for nanomaterial characterization is transmission electron microscopy (TEM). TEM allows for the visualization of nanomaterials with atomic-scale resolution by transmitting a beam of electrons through a thin sample. The interaction between the electron beam and the material generates contrast and diffraction patterns, which provide information about the morphology, size, and crystallographic structure of nanoparticles. Advanced TEM techniques, such as high-resolution TEM (HRTEM) and scanning transmission electron microscopy (STEM), extend these capabilities by offering detailed insights into lattice fringes and atomic arrangements, essential for studying defects and heterostructures.

X-ray diffraction (XRD) is another indispensable technique that provides crucial information about the crystalline structure of nanomaterials. By directing X-rays onto a sample and measuring the diffraction pattern, XRD can determine lattice parameters, crystallite size, and phase identification. For nanomaterials, line broadening in XRD peaks due to finite crystallite sizes can be analyzed using the Scherrer equation to estimate particle size. Moreover, advanced methods like synchrotron-based XRD and *in situ* XRD studies enable real-time monitoring of structural changes under varying conditions, facilitating the investigation of dynamic processes such as phase transitions and chemical reactions.

Surface analysis techniques such as X-ray photoelectron spectroscopy (XPS) and Auger electron spectroscopy (AES) offer detailed insights into the chemical composition and electronic states of nanomaterial surfaces. XPS, in particular, is a powerful tool for determining elemental composition, oxidation states, and the presence of functional groups, crucial for understanding surface chemistry and interactions. By analyzing the binding energy of photoelectrons emitted from the surface atoms, XPS elucidates the chemical environment and electronic states, aiding in the assessment of surface modifications or functionalization.

Scanning electron microscopy (SEM) complements TEM by providing high-resolution imaging of nanomaterial morphology, albeit with lower magnification capabilities. SEM is particularly useful for examining surface topogra-

10.5. CHARACTERIZATION TECHNIQUES FOR NANOMATERIALS

phies, particle shape, and aggregates, utilizing a focused beam of electrons that interacts with the sample surface to generate secondary electrons, backscattered electrons, or characteristic X-rays. Advanced SEM techniques, such as environmental SEM (ESEM) and field emission SEM (FESEM), enable imaging under varied environmental conditions and offer enhanced resolution, respectively, allowing the examination of hydrated or beam-sensitive nanomaterials.

Atomic force microscopy (AFM) is a versatile technique that provides three-dimensional surface profiles with nanometer-scale resolution. Unlike electron microscopy, AFM uses a sharp tip mounted on a cantilever that scans over the sample surface, interacting with intermolecular forces. AFM not only delivers topographic information but also provides mechanical, electrical, and magnetic property data through modes like tapping mode, contact mode, and force spectroscopy. This adaptability makes AFM particularly valuable for characterizing soft or biological nanomaterials and for performing nanoscale manipulations or modifications.

Raman spectroscopy is an optical characterization technique that probes molecular vibrations and provides insights into the chemical composition and crystallinity of nanomaterials. By measuring the inelastic scattering of monochromatic light, Raman spectroscopy identifies characteristic vibrational modes, allowing the identification of chemical structures and phases. This technique is especially useful for characterizing carbon-based nanomaterials such as graphene and carbon nanotubes, where distinct Raman signatures correlate with structural defects, doping levels, and electronic properties.

Dynamic light scattering (DLS) is a technique widely employed to determine the hydrodynamic size distribution of nanoparticles in suspension. DLS measures fluctuations in the intensity of scattered light due to Brownian motion of particles, providing estimates of particle size and polydispersity. It is extensively used in colloidal science and bio-nanotechnology for analyzing nanoparticle aggregation and stability in various media, contributing to the development of formulations with controlled particle distributions for drug delivery and therapeutic applications.

For the characterization of optical properties, ultraviolet-visible (UV-Vis) spectroscopy offers quantitative data on light absorption and transmission characteristics, which are influenced by the size, shape, and composition of nanomaterials. UV-Vis spectroscopy is critical in determining the band gap of semiconductor nanoparticles and in monitoring the synthesis of noble metal colloids, where localized surface plasmon resonances manifest as distinct absorption bands.

Infrared (IR) spectroscopy is another vital tool for analyzing the vibrational modes of molecular bonds present in nanomaterials. IR spectroscopy identifies functional groups on nanomaterial surfaces and assesses chemical modifications, making it a cornerstone method for characterizing organic-inorganic hybrid systems and polymer nanocomposites. Additionally, Fourier-transform infrared (FTIR) spectroscopy enhances sensitivity and resolution, enabling detailed structural information about complex nanostructures.

In addition to these techniques, mass spectrometry, particularly matrix-assisted laser desorption/ionization (MALDI) and time-of-flight (TOF) mass spectrometry, offers complementary data on nanomaterial composition and structure. These methods are particularly advantageous in analyzing large biomolecules or polymer-based nanomaterials, where they provide information on molecular weight distribution, fragmentation patterns, and copolymer compositions.

Each characterization technique brings a distinct perspective, underscoring the importance of a multifaceted approach in nanomaterial analysis. By integrating complementary techniques, researchers can obtain a comprehensive understanding of the material's complete profile, elucidating the interconnected relationships between structure, composition, and function.

The role of computational methods in conjunction with experimental characterization has grown increasingly significant. Techniques such as density functional theory (DFT) simulation, molecular dynamics (MD), and ab-initio calculations allow for the prediction and interpretation of electronic structures, thermodynamical properties, and reaction mechanisms at an atomic level. These computational approaches provide a theoretical framework that supports the experimental findings and offers predictive insights into unexplored nanomaterials, accelerating the discovery process.

Ensuring reliability and reproducibility in nanomaterial characterization is critical yet challenging, given the sensitivity of nanomaterials to preparation methods and environmental factors. Standardization of techniques and rigorous quality control procedures are essential to obtain consistent and comparable data across research settings, fostering advancements in the field.

Characterization techniques for nanomaterials are indispensable tools that drive the continuous innovation and understanding of materials at the nanoscale. By harnessing their power, scientists can unlock the full potential of nanomaterials, transforming discoveries into transformative technologies across diverse domains, from healthcare to energy and beyond. These techniques collectively underpin the progress and promise of nanotechnology, guiding researchers in the ongoing quest to engineer functional materials from the ground up.

10.6 Applications of Nanotechnology in Industry

Nanotechnology represents a transformative force across various industrial sectors by enabling the manipulation of materials at the atomic and molecular scale. This profound capability allows for the development of products and processes with enhanced performance, new functionalities, and improved efficiencies. The following examination explores how nanotechnology's principles and methodologies extend into key industries, offering innovative solutions and driving unprecedented advancements.

In the electronics industry, nanotechnology is at the forefront of miniaturizing devices, reducing power consumption, and enhancing performance. The ongoing evolution of integrated circuits follows Moore's Law, transitioning to nodes as small as 5 nm with the aid of nanolithography techniques such as extreme ultraviolet (EUV) lithography. Carbon-based nanomaterials, such as graphene and carbon nanotubes, are being explored as potential successors to silicon due to their superior electrical properties and the potential for faster, more efficient conductors. Their application extends to flexible electronics, wearables, and advanced display technologies, contributing to the development of ultra-thin, lightweight, and transparent electronic components.

In energy, nanotechnology provides significant breakthroughs in both storage and conversion technologies. Lithium-ion batteries, crucial for the proliferation of electric vehicles and renewable energy storage, witness immense improvements through nanostructured materials utilized as electrodes. These materials offer increased surface area, enhanced electron transport, and more efficient ion diffusion pathways, resulting in higher energy capacities and faster charging times. Innovations in nanostructured silicon or tin-based anodes promise to extend battery lifespan and capacity, overcoming limitations faced by traditional graphite-based systems.

Solar energy conversion benefits from nanotechnology through the development of thin-film photovoltaic cells and perovskite solar cells, which offer higher efficiencies and lower manufacturing costs compared to conventional silicon solar cells. The incorporation of quantum dots, with their size-tunable band gaps, enhances light absorption efficiency and, consequently, the power conversion efficiency of photovoltaic devices. Furthermore, nanotechnology facilitates the creation of advanced coatings and nanostructures that minimize reflection and increase light trapping, further optimizing solar energy capture.

In healthcare and medicine, nanotechnology fosters breakthroughs in diagnostics, therapy, and drug delivery systems. Nanoparticles serve as carriers that can target specific cellular sites, enabling the precise delivery of therapeutic agents while minimizing side effects. Liposomes, dendrimers, and polymeric nanoparticles exemplify nanoscale carriers used in the controlled release of drugs, anti-cancer agents, and gene therapy vectors. Additionally, nanotechnology enhances imaging techniques, with nanoparticles functioning as contrast agents in magnetic resonance imaging (MRI), computed tomography (CT), and fluorescent imaging, allowing for earlier and more accurate disease detection.

Biosensors leverage nanotechnology to achieve heightened sensitivity and specificity in detecting biomolecules, pathogens, or environmental contaminants. Nano-scaled biosensors utilize materials such as gold nanoparticles and carbon nanostructures to detect low concentrations of analytes in complex mixtures, opening opportunities for personalized medicine, and real-time health condition monitoring.

The materials science sector harnesses nanotechnology to develop nanocomposites and coatings with superior mechanical, thermal, and chemical properties. By incorporating nanoparticles such as carbon nanotubes, graphene oxide, or metal oxides into a polymer matrix, manufacturers produce materials with enhanced strength, modulus, and durability for use in the automotive, aerospace, and construction industries. Nano-engineered coatings enhance corrosion resistance, thermal insulation, and hydrophobicity, extending the lifespan of materials exposed to harsh environments.

In the environmental sector, nanotechnology provides innovative solutions for pollution reduction, water purification, and resource management. Nanomaterials designed for catalytic applications enable cleaner production processes and more effective decomposition of pollutants. For instance, titanium dioxide nanoparticles in catalytic converters facilitate the breakdown of vehicular emissions into less harmful substances, while magnetic nanoparticles assist in removing heavy metal contaminants from wastewater. Additionally, nanotech-

10.6. APPLICATIONS OF NANOTECHNOLOGY IN INDUSTRY

nology advances water purification systems, with nanofiltration membranes enhancing the removal efficiency of pathogens and toxins from drinking water supplies.

Agriculture and food industries increasingly incorporate nanotechnology to improve crop yields, food safety, and packaging. Nanomaterials such as nanoscale fertilizers, pesticides, and delivery systems enable the controlled release of nutrients and agrochemicals, reducing waste and environmental impact. In food safety, nanotechnology offers innovative solutions for pathogen detection and food preservation. Intelligent packaging systems incorporate nanosensors to monitor the freshness and condition of food products, while nano-coatings extend shelf life by preventing microbial growth.

Nanotechnology is also pivotal in textile and consumer goods industries, where functionalities such as stain resistance, antibacterial properties, and UV protection are embedded into fabrics through nano-finishing processes. Nanotextiles combine comfort, durability, and multi-functionality, catering to the evolving needs of consumer markets and providing smart textiles capable of monitoring physiological parameters or adapting to environmental changes.

Beyond its current applications, nanotechnology's role in advancing industrial processes continues to expand with ongoing research into self-healing materials, neuromorphic computing, and quantum technologies. Self-healing materials developed through the incorporation of capsule-containing nanoparticles possess the ability to repair themselves autonomously following damage, promising applications in smart coatings and construction materials.

Neuromorphic computing seeks to replicate the function of neural systems using nanomaterials such as memristors and RRAM (resistive random-access memory), which simulate synaptic activity, offering avenues for energy-efficient data processing and artificial intelligence.

Quantum computing, poised to revolutionize information processing through the exploitation of quantum mechanical phenomena, is another frontier propelled by nanotechnology advances. Fabrication of qubits from superconducting nanostructures, topological insulators, or trapped ions necessitates precise control at the atomic scale, underscoring nanotechnology's critical role in realizing the next generation of computational technology.

The applications of nanotechnology within industry span every facet of modern life, continuously redefining possibilities and setting new standards for innovation. The confluence of material properties at the nanoscale with cutting-edge synthesis, fabrication, and characterization techniques heralds a future where nanotechnology is intricately interwoven into the fabric of society to address the complex challenges of sustainability, health, and technological progress. The ongoing synergy between research, development, and commercial deployment ensures that nanotechnology remains a cornerstone of industrial advancement, poised to unlock its full potential for generations to come.

10.7 Environmental and Ethical Implications of Nanotechnology

The emergence of nanotechnology has ushered in a paradigm shift in various fields, bringing forth promises of unprecedented advancements in technology, medicine, and environmental management. However, with these revolutionary capabilities come significant considerations regarding their environmental and ethical implications. As nanotechnology continues to evolve, it is critical to evaluate both its potential impacts on ecological systems and the ethical foundations guiding its development and application.

One of the cornerstone environmental implications of nanotechnology revolves around its production processes and lifecycle. The synthesis of nanomaterials often necessitates the use of chemicals and solvents that may pose significant environmental hazards if not properly managed. For instance, emissions from chemical vapor deposition (CVD) or the waste products generated from sol-gel processes could contribute to pollution unless stringent disposal and treatment protocols are implemented. Nanotechnology's production and application phases require comprehensive environmental impact assessments to predict and mitigate potential adverse effects on surrounding ecosystems.

Moreover, the widespread incorporation of nanoparticles into consumer products has raised concerns about their release into the environment. The unique reactivity and catalytic properties of nanoparticles, while beneficial in technological applications, can lead to unforeseen effects when these particles enter natural ecosystems. One specific concern is the potential for nanoparticles to interact with biological organisms, causing harmful effects not typical of their bulk counterparts. For instance, silver nanoparticles, widely used for their antimicrobial properties in medical and consumer products, can disrupt microbial communities in aquatic systems, affecting water quality and biodiversity.

10.7. ENVIRONMENTAL AND ETHICAL IMPLICATIONS OF NANOTECHNOLOGY

The behavior and fate of nanomaterials in natural environments represent critical research areas. Factors such as aggregation, dissolution, and surface modification influence their environmental mobility and potential toxicity. Studying these nanoparticles' interactions with environmental components, like soil, water, and air, is essential to understanding and managing their ecological footprint. Some nanoparticles may accumulate in organisms along the food chain, resulting in bioaccumulation and potentially posing risks to human health and animal populations.

Furthermore, nanoparticles' ability to undergo transformations through processes like oxidation or reduction in environmental settings can alter their behavior and toxicity. For example, the transformation of carbon-based nanomaterials, such as fullerenes and graphene oxides, may result in the formation of toxic by-products that differ significantly from their original formulations. These transformations necessitate detailed environmental monitoring and the development of guidelines to understand and control their ecological implications.

The ethical implications of nanotechnology development and implementation are equally significant and multi-faceted. These implications are centered on the principles of responsibility, equity, and transparency in research and development processes. The rapid growth of nanotechnology demands careful consideration of its socioeconomic impacts and the potential for exacerbating existing inequalities. As industrialized nations lead the charge in nanotechnology advancements, there is a risk that developing countries may be left behind or exploited, exacerbating global disparities.

Ethical concerns also arise from the potential misuse or dual-use nature of nanotechnology. The ability to engineer materials at the atomic scale opens possibilities for military applications and surveillance technologies, raising questions about their regulation and control. Public engagement and policy frameworks are vital components in addressing these concerns, ensuring that the development of nanotechnology aligns with societal values and priorities.

Intellectual property rights and patenting practices present additional ethical challenges in the field of nanotechnology. The dominance of a few large corporations could stifle innovation and limit access to nanotechnologies, particularly in less affluent regions. Encouraging open innovation and partnerships between industry, academia, and government may foster more equitable access to nanotechnology advancements and benefits.

Regulatory frameworks and international agreements are critical to addressing these multifaceted ethical and environmental challenges. The establishment of clear guidelines for nanomaterial synthesis, handling, disposal, and labeling can minimize risks associated with accidental exposure or environmental release. However, the dynamic nature of nanotechnology, with its rapid evolution and diverse applications, poses challenges for regulators seeking to strike a balance between fostering innovation and ensuring safety.

Public perception and trust in nanotechnology also play a pivotal role in shaping its trajectory. Transparent communication of risks and benefits, along with active public participation in decision-making, can cultivate a more informed and engaged public. Such engagement enhances the legitimacy and acceptance of policies and practices surrounding nanotechnology, ensuring its development aligns with the broader public interest.

Ethics education and interdisciplinary approaches play a crucial role in fostering a holistic understanding of nanotechnology's implications and guiding responsible research and development. Institutions and educators must emphasize the integration of ethical considerations into science and engineering curricula, equipping future researchers and practitioners with the tools to recognize and address moral challenges in their work.

Nanotechnology's potential to address pressing global challenges, such as climate change and sustainable development, must be balanced with a commitment to environmental stewardship and ethical integrity. By minimizing waste, reducing resource consumption, and enabling the creation of cleaner technologies, nanotechnology can significantly contribute to achieving sustainability goals. Efforts to engineer nanoscale solutions for renewable energy generation, water purification, and pollution remediation represent hopeful strides toward a more sustainable future.

The environmental and ethical implications of nanotechnology represent critical areas of discourse that necessitate a proactive and informed approach. By acknowledging potential risks and fostering collaborative frameworks that prioritize ethical considerations, the nanotechnology community can ensure that scientific and technological advancements serve the collective good. The path forward must be characterized by vigilance, dialogue, and collective responsibility, ensuring that nanotechnology fulfills its promise as a tool for progress while safeguarding the integrity of natural ecosystems and societal values.

Chapter 11

Biochemical Thermodynamics

Biochemical thermodynamics applies the laws of thermodynamics to the energetic transformations within biological systems, offering a quantitative perspective on life processes. This chapter explores the role of Gibbs free energy in biochemical reactions, elucidating how living organisms drive non-spontaneous processes by coupling them with spontaneous ones. The thermodynamic analysis of enzyme catalysis and metabolic pathways reveals the intricate balance between energy input and output critical for cellular function. By examining binding interactions and membrane transport, the chapter provides insights into how biological systems maintain homeostasis and adapt to changing environments. These principles underscore the energetic efficiency and regulatory mechanisms that are fundamental to bioenergetics and metabolic control.

11.1 Principles of Biochemical Thermodynamics

Biochemical thermodynamics serves as a critical framework for understanding the energetic transformations that occur within biological systems. This field is anchored on the principles of classical thermodynamics, with an emphasis on their applicability to the complex and dynamic environment of biosystems. The fundamental laws governing energy transformations can elucidate how living organisms maintain order and carry out various functions essential for survival.

The first and second laws of thermodynamics provide the foundation for analyzing biochemical processes. The first law, the law of energy conservation, asserts that energy cannot be created or destroyed in an isolated system, only transformed from one form to another. This principle is critical when considering energy exchanges within living cells, which must adhere to this conservation law even as they engage in various biochemical reactions.

A rigorous analysis of energy transformations involves understanding various thermodynamic quantities. Enthalpy (H), a measure of total energy, includes both internal energy (U) and the product of pressure and volume (PV):

$$H = U + PV$$

Within biological systems, changes in enthalpy (ΔH) are indicative of heat exchange during constant pressure processes, which is often the context within cells. Such changes provide insight into whether reactions absorb heat (endothermic) or release heat (exothermic).

While the first law addresses the conservation of energy, the second law of thermodynamics introduces the concept of entropy (S), positing that in any spontaneous process, the total entropy of the system and its surroundings always increases. Entropy can be understood as a measure of randomness or disorder within a system. In biochemical systems, the drive towards increased entropy is counterbalanced by highly organized processes that maintain low entropy locally, a hallmark of living organisms.

The ability of a system to do work is gauged by Gibbs free energy (G), a central concept in understanding biochemical reactions:

$$G = H - TS$$

where T is the absolute temperature in Kelvin. A change in Gibbs free energy (ΔG) predicts the direction and spontaneity of a chemical reaction under constant temperature and pressure. A negative ΔG indicates a spontaneous process, while a positive ΔG suggests non-spontaneity:

$$\Delta G = \Delta H - T\Delta S$$

11.1. PRINCIPLES OF BIOCHEMICAL THERMODYNAMICS

In biochemical contexts, the practicality of ΔG lies in its ability to predict reaction feasibility. This is critical when considering metabolic pathways, where certain reaction steps must proceed to sustain life, despite unfavorable energy dynamics.

Reactions are often coupled to allow non-spontaneous processes to occur by harnessing the energy from spontaneous ones, a strategy ubiquitously employed by cells. For example, the synthesis of adenosine triphosphate (ATP) is driven by energy released from the catabolism of glucose. The hydrolysis of ATP, typically an exergonic reaction with a $\Delta G^0 \approx -30.5$ kJ/mol, serves as an energy currency that powers countless endergonic processes within the cell.

Understanding the equilibrium state of biochemical reactions is further facilitated by the application of the laws of thermodynamics. At equilibrium, the rate of the forward reaction equals that of the reverse reaction, and the Gibbs free energy change (ΔG) is zero. The relation between ΔG and the equilibrium constant (K_{eq}) of a reaction is described by the equation:

$$\Delta G^0 = -RT \ln K_{eq}$$

where R is the universal gas constant and T is the temperature in Kelvin. This equation links thermodynamic principles to the chemical equilibrium, allowing the quantification of reaction tendencies based on changes in concentration.

Biological systems constantly drive reactions away from equilibrium through mechanisms such as substrate cycling and feedback inhibition, maintaining a dynamic equilibrium essential for cellular functionality. These mechanisms are finely tuned, enabling cellular adaptation to varying environmental conditions.

The pursuit of lower Gibbs free energy states governs the directionality of metabolic pathways. The dynamic nature of such systems necessitates that we consider actual cellular conditions rather than standard conditions (ΔG^0), transforming our analysis to account for concentrations of substrates and products, temperature variances, and cellular environment peculiarities.

Moreover, thermodynamic principles extend to the structure and behavior of macromolecules. Proteins, nucleic acids, and lipids exhibit unique energetic profiles in response to their respective environments. The stability of protein structures, for instance, is often dictated by enthalpic interactions such as hydrogen bonds, ionic interactions, and van der Waals forces, calibrated by entropic considerations like conformational flexibility and solvent effects.

In biomolecular folding, the balance between enthalpy and entropy shifts as proteins transition from unfolded to folded states. Initial folding involves a loss of entropy due to decreased conformational freedom, counteracted by the formation of enthalpically favorable intramolecular interactions. The overall thermodynamic stability of the folded protein is crucial for its biological function, highlighting a critical area where thermodynamic analysis is essential.

By employing the thermodynamic framework, researchers can elucidate pathways and mechanisms underpinning phenomena such as enzyme catalysis, signal transduction, and the thermoregulation of biological membranes. Studying how biological membranes facilitate energy transformation, for instance, reveals insights into ion gradients and membrane potential maintenance, which are thermodynamically essential for processes like cellular respiration and photosynthesis.

Principles of biochemical thermodynamics provide a systematic approach to quantify and predict cellular processes. The complex interplay of energy transformations underlies the ability of living organisms to sustain life, from which we can derive comprehensive insights into cell biology, molecular physiology, and the broader fields of bioenergetics and metabolic control. The delicate balance of energy conservation, entropy modulation, and Gibbs free energy management within cells reflects the sophistication of life driven by thermodynamic principles.

11.2 Free Energy in Biological Reactions

In biochemical systems, the concept of free energy is pivotal in understanding the spontaneity and directionality of reactions. Gibbs free energy (G) serves as a reliable indicator of whether a reaction can occur spontaneously under given conditions. As defined previously, the expression for Gibbs free energy is given by:

$$G = H - TS$$

where H is enthalpy, T is temperature in Kelvin, and S is entropy. Changes in Gibbs free energy (ΔG) provide insight into the feasibility of biochemical reactions. The conditions under which this evaluation takes place typically

11.2. FREE ENERGY IN BIOLOGICAL REACTIONS

include constant temperature and pressure, characteristics of most biological systems.

Analyzing ΔG, one can predict reaction spontaneity:

- $\Delta G < 0$: The reaction is exergonic and can proceed spontaneously.
- $\Delta G = 0$: The system is at equilibrium; no net change will occur.
- $\Delta G > 0$: The reaction is endergonic and non-spontaneous under the current conditions.

The sign and magnitude of ΔG are influenced by both the environmental conditions and the concentrations of reactants and products. Furthermore, the standard Gibbs free energy change (ΔG^0), a constant associated with reactions under standard conditions, is defined at pH 7 in biological systems and provides a reference point for comparing reaction tendencies:

$$\Delta G = \Delta G^0 + RT \ln \left(\frac{[C]^c [D]^d}{[A]^a [B]^b} \right)$$

Here, R is the universal gas constant, and the terms within the parentheses represent the reaction quotient, the ratio of product and reactant concentrations, each raised to the power corresponding to their stoichiometric coefficients.

In cellular environments, reactions rarely achieve the conditions defined by ΔG^0, yet understanding this value allows scientists to infer the changes that actual concentration alterations might impart on free energy. Consequently, ΔG is more reflective of the in vivo state:

$$\Delta G = \Delta G^0 + RT \ln Q$$

where Q is the reaction quotient. This formulation underscores the adaptability of biochemical pathways in responding to varying cellular demands, maintaining homeostasis despite fluctuating environmental factors.

Many metabolic reactions require an input of energy to proceed. This energy is often provided through coupling with exergonic reactions, a prevalent cellular mechanism that fosters reactions which, by their nature, would not be spontaneous. A quintessential example is the synthesis of macromolecules, requiring substantial energy input typically derived from ATP hydrolysis:

$$\text{ATP} + \text{H}_2\text{O} \rightarrow \text{ADP} + \text{P}_i + \text{Energy}$$

ATP (adenosine triphosphate), upon hydrolysis to ADP (adenosine diphosphate) and inorganic phosphate (P_i), releases about -30.5 kJ/mol of free energy under standard conditions. This highly exergonic process is capable of driving many unfavorable biochemical transformations.

Beyond the synthesis of biomolecules, ATP hydrolysis provides energy for processes ranging from active transport across membranes to the mechanical work of muscle contraction. Cellular respiration, the process by which cells extract energy from glucose through a series of oxidative reactions, capitalizes on this principle by generating a proton gradient across mitochondrial membranes, a crucial component in ATP synthesis.

In terms of oxidative phosphorylation, the transformation of chemical energy into ATP within mitochondria involves electron transport chains (ETCs) energizing proton pumps, which relocate protons across the inner mitochondrial membrane to establish a gradient. This chemiosmotic potential facilitates ATP synthesis via ATP synthase, showcasing the intricate coupling between chemical gradients and ATP generation.

The role of Gibbs free energy is further exemplified in the glycolytic pathway, where intermediate steps are characterized by specific changes in free energy. As glycolysis progresses through a series of ten enzymatic reactions, the end products are two molecules of pyruvate, further illustrating how coupled reactions can manage overall pathway energy dynamics. Herein, reactions such as the phosphorylation of glucose and its conversion to fructose-1,6-bisphosphate require ATP, whereas subsequent steps liberate energy.

Enzymatic catalysis, in particular, is deeply intertwined with thermodynamic principles. Enzymes operate by lowering the activation energy barrier, expediting reactions that would otherwise progress at imperceptibly slow rates under physiological conditions. The role of enzymes in controlling $\Delta G^\ddagger$, the Gibbs free energy of activation, is critical for managing biochemical reaction rates, thereby influencing pathway kinetics and cellular energy efficiency.

A central facet of free energy analysis in biology is the effect of environmental changes, such as pH and temperature, on reaction dynamics. Cellular environments maintain specific pH levels optimal for enzymatic activity through buffers and compartmentalization. Temperature, a critical influencing factor, modulates the kinetic energy of molecules, directly impacting the rates of reaction and thus the Gibbs free energy change. This adaptability underscores the sensitivity of living systems to energetic and thermodynamic conditions.

Biochemists also explore the implications of free energy changes in the context of signal transduction pathways, where small molecules or ions convey information by binding to receptor proteins, instigating conformational alterations that trigger downstream signaling events. Understanding the energetic basis allows for predictions regarding signaling efficiencies and potential disruption impacts, such as during dysregulated intracellular communication.

Drilling deeper into free energy concepts, the examination of reaction coupling and pathways enables targeted intervention strategies in biotechnology and medicine. Enzyme inhibitors, allosteric modulators, or compartment-specific alteration of concentration gradients, upon robust thermodynamic analyses, advance the development of therapeutic modalities and metabolic engineering to optimize production pathways for desired substances.

Gibbs free energy not only formulates the theoretical groundwork for considering reaction spontaneity but also bridges biochemical pathway regulation and cellular function. Its principles unveil an empirical spectrum through which biological reactions may be quantitatively assessed, revealing insights into the biochemical choreography that sustains life.

This nuanced understanding of free energy extends its relevance to fields beyond biology, embodying universal laws that clarify molecular processes across a spectrum of life forms, maintaining the delicate balance of order and chaos inherent in living systems.

11.3 Enzyme Thermodynamics and Kinetics

The intersection of thermodynamics and kinetics is pivotal in understanding enzymatic reactions, revealing insights into both the energetic landscapes and the rates at which biochemical reactions occur. Enzymes, as biological catalysts, exemplify extraordinary specificity and efficiency, owing much of their functionality to the interplay between thermodynamic principles and kinetic constraints. This section delves into the sophisticated coordination of these two domains, explaining how enzymes lower activation energy, influence reaction dynamics, and integrate into metabolic pathways.

At the heart of enzyme activity is the concept of activation energy (E_a), which represents the minimum energy required to convert reactants into products.

Enzymes facilitate reactions by reducing E_a, thereby increasing the reaction rate without altering the overall ΔG of the reaction. This process is integral in ensuring that reactions necessary for life occur at perceptible rates under physiological conditions.

The transition state theory provides a theoretical framework for understanding how enzymes boost reaction rates. According to this theory, the transition state represents a high-energy, unstable arrangement of atoms that can decay to form products. Enzymes stabilize this transition state through specific active site interactions, lowering the activation energy required. The thermodynamic snapshot of this phenomenon is illustrated by comparing energy profiles with and without enzymatic catalysis.

$$\Delta G^{\ddagger} = G_{\text{transition state}} - G_{\text{reactants}}$$

This equation highlights the reduction in $\Delta G^{\ddagger}$ attributable to enzymatic intervention. The formation of an enzyme-substrate complex (ES) further elucidates this dynamic, following the reaction sequence:

$$E + S \rightleftharpoons ES \rightarrow E + P$$

where E denotes the enzyme, S the substrate, and P the product.

Enzyme kinetics, primarily governed by the Michaelis–Menten model, provides quantitative insights into the relationship between reaction velocity and substrate concentration. The model posits the formation of the enzyme-substrate complex as a critical intermediate:

$$v = \frac{V_{\max}[S]}{K_m + [S]}$$

Here, v is the rate of reaction, $V_{\max}$ the maximum reaction velocity, and K_m the Michaelis constant—a reflection of substrate affinity. A small K_m indicates high affinity, whereas larger values suggest weaker binding.

The intricacies of enzyme function are communicated through these parameters, with $V_{\max}$ directly proportional to enzyme concentration and number of active sites available, while K_m offers insights into the binding efficiency and interaction dynamics within the active site.

11.3. ENZYME THERMODYNAMICS AND KINETICS

Enzyme activity is heavily influenced by thermodynamic parameters such as temperature and pH, both of which can alter enzyme conformation and functionality. According to the Arrhenius equation, the rate constant k changes with temperature:

$$k = Ae^{-E_a/RT}$$

where A is the pre-exponential factor, and R is the universal gas constant. This relationship underscores the concept of temperature optima; enzymes often have an optimal temperature where catalytic activity peaks before thermal denaturation disrupts their structure.

Enzyme regulation, through mechanisms such as allosteric control and covalent modification, emphasizes kinetic modulation rooted in thermodynamic principles. Allosteric enzymes do not follow Michaelis–Menten kinetics, as the presence of allosteric sites allows for cooperative binding and conformational changes influencing activity. The thermodynamic basis for such regulation is entrenched in the structure-function relationship, where ligand binding induces a shift in energy landscapes, stabilizing either active or inactive enzyme conformations.

Enzymatic pathways further integrate kinetic and thermodynamic principles to achieve fine-tuned metabolic control. A classic case study involves feedback inhibition in metabolic pathways where end products inhibit earlier steps by binding to allosteric sites, thereby regulating pathway flux based on both substrate availability and cellular demand. This regulatory architecture ensures that energy and resource utilization remain efficient and responsive.

Exploring enzymatic catalysis through a thermodynamic lens also accentuates the energy landscapes that dictate substrate binding and transformation. Conformational selection and the induced fit model underscore flexible interactions between enzymes and substrates, harmonizing structural and energetic compatibilities that facilitate catalysis.

Consider the role of enzymes in the digestion of macromolecules. Proteases catalyze the hydrolysis of peptide bonds, an energetically attractive yet kinetically hindered process. By forming transient covalent intermediates and polarizing peptide carbonyl groups, proteases demonstrate the enzymatic proficiency in overcoming kinetic barriers while conserving free energy profiles.

Enzymomes, representing large ensembles of enzymes functioning collectively, underscore the complexity of enzymatic networks wherein kinetic and thermodynamic integration is paramount for metabolic homeostasis. Within the purview of systems biology, the study of enzymomes explores synergistic interactions and modularity, shedding new light on enzyme evolution and adaptation.

Additionally, advancements in computational biology have leveraged molecular dynamics simulations to visualize enzyme catalysis, providing atomistic insights into the catalytic cycles and energy barriers involved in enzymatic reactions. Such investigations afford molecular-level accuracy in elucidating the dynamics of enzyme-substrate complexes and transition states, extending our comprehension of enzymatic function beyond static models.

The role enzymes play in facilitating life-sustaining reactions reverberates across evolution, wherein evolutionary pressures have honed enzyme efficiency and specificity, adapting them exquisitely to their respective cellular niches and environmental conditions. Through a fusion of kinetic and thermodynamic principles, enzyme catalysis continues to be an area of intense study, contributing to innovations in biotechnology, drug development, and synthetic biology.

The intertwined nature of enzyme thermodynamics and kinetics affords a comprehensive understanding of how enzymes act as efficient catalysts of life processes. This synthesis of foundational principles reveals the intricate coordination necessary for maintaining biological order and function, providing a gateway to further exploring biological catalysis and engineering novel enzymatic capabilities. Enzymes embody the quintessential embodiment of precision and efficacy, stemming from their adept manipulation of energy landscapes and reaction kinetics.

11.4 Thermodynamics of Metabolic Pathways

The thermodynamics of metabolic pathways underpins a vast array of biochemical transformations that sustain life, orchestrating the intricate web of metabolism within a cell. These pathways function as networks of enzyme-catalyzed reactions, each step governed by thermodynamic principles that dictate directionality and energetics. This section delves into the principles and applications of thermodynamics in understanding key metabolic pathways, focusing on the flow of energy, the concept of coupled reactions, and the maintenance of cellular homeostasis.

11.4. THERMODYNAMICS OF METABOLIC PATHWAYS

Metabolic pathways are categorized into anabolism and catabolism. Anabolic pathways are responsible for biosynthetic processes, building complex molecules from simpler ones and generally require energy input. Catabolic pathways involve the breakdown of complex molecules into simpler units, releasing energy harnessed for cellular work. The interplay between these two forms an integrated metabolic system known as metabolism, with thermodynamics guiding the energetic feasibility and control of these pathways.

At the core of metabolic thermodynamics is the concept of free energy change (ΔG), which provides insight into the spontaneous nature of reactions within pathways. While individual steps in a pathway can be endergonic (non-spontaneous with $\Delta G > 0$), the overall pathway can be exergonic (spontaneous with $\Delta G < 0$) due to coupling of reactions. This coupling allows energy-releasing reactions to drive energy-requiring ones, a fundamental principle that enhances metabolic efficiency.

A quintessential example is glycolysis, the biochemical pathway that metabolizes glucose to pyruvate, generating ATP and NADH in the process. Despite involving both exergonic and endergonic steps, the pathway as a whole is exergonic with a net release of free energy, driving glycolysis forward in the cellular environment:

- **Phosphorylation of Glucose:** The initial step involves the phosphorylation of glucose to glucose-6-phosphate, an endergonic reaction driven by the exergonic hydrolysis of ATP. Herein, reaction coupling is evident:

$$\text{Glucose} + \text{ATP} \rightarrow \text{Glucose-6-Phosphate} + \text{ADP}$$

- **Formation of Pyruvate:** The final steps of glycolysis, involving the conversion of phosphoenolpyruvate to pyruvate, are highly exergonic, liberating energy captured through substrate-level phosphorylation to produce ATP.

Coupled reactions are further exemplified in cellular respiration during oxidative phosphorylation. This process describes the coordinated activity of the electron transport chain (ETC) and ATP synthase within mitochondria. Electrons transferred through protein complexes of the ETC create a proton gradient across the inner mitochondrial membrane. This electrochemical gradient, a form of potential energy, is then utilized by ATP synthase to convert

ADP and inorganic phosphate into ATP, an elegant demonstration of energy conversion governed by chemiosmotic principles.

The thermodynamic favorability of ATP production hinges on well-maintained gradients and efficient coupling mechanisms, reflecting the importance of compartmentalization and membrane integrity in cellular energetics.

In examining metabolic pathways, it is crucial to consider the concept of metabolic flow and equilibrium. Metabolic pathways do not reach true thermodynamic equilibrium, as this would halt metabolic activity. Instead, cells maintain a dynamic steady state or homeostasis through continuous flux, sustained by thermodynamic driving forces and feedback regulation.

Metabolic flexibilities, such as the regulation of pathway intermediates and the allosteric regulation of key enzymes, control pathway flux and prevent the accumulation of intermediates. One classic example is the regulation of the citric acid cycle, where key enzymes like isocitrate dehydrogenase and α-ketoglutarate dehydrogenase are modulated by the energy charge of the cell, receiving inputs regarding ATP availability and NADH/NAD$^+$ ratios.

Enzymatic controls allow pathways to adapt, turning precise thermodynamic principles into practical regulations that reflect cellular needs and environmental signals. The pathways integrate reciprocal control mechanisms, ensuring that anabolic and catabolic pathways operate in a coordinated fashion, preventing futile cycles that would otherwise lead to wasted energy.

Switching focus to anabolic pathways, the synthesis of fatty acids is a prime illustration of thermodynamic investments. Fatty acid synthesis from acetyl-CoA requires significant ATP equivalents and reducing power in the form of NADPH, sourced from the pentose phosphate pathway. The commitment of resources is justified by the thermodynamic stability and storage capability of fatty acids and triglycerides, highlighting the interplay between energy storage forms, conversion, and utilization.

In the context of thermodynamics, metabolic pathways must adapt to variabilities such as substrate availability, demand for products, and environmental stresses. Allosteric interactions, covalent modifications, and hormonal controls, such as insulin and glucagon, further integrate metabolic pathways with systemic regulation, ensuring metabolic networks respond appropriately to maintain organismal homeostasis.

Moreover, the advent of system-level analysis tools, including metabolic engineering and computational modeling, affords the potential to quantify thermodynamic parameters across whole-cell networks, revealing insights into emergent properties and global control strategies. Metabolic flux analysis, for instance, utilizes isotopic labeling and mass spectrometry to determine pathway efficiencies and bottlenecks, providing a microscopic view of macromolecular thermodynamics that guide pathway interventions.

Thermodynamic considerations also extend to ecological and evolutionary contexts, where energy efficiencies influence organismal fitness and adaptability. Evolutionary pressures favoring energy-conserving strategies are evident in adaptive enzyme coding, pathway redundancies, and cross-linkages, emerging over time as robust, flexible, and resilient metabolic repertoires.

The thermodynamics of metabolic pathways forms the bedrock understanding of cellular biology, enabling life to harness, convert, and utilize energy through an exquisitely coordinated interplay of chemical reactions. These pathways manifest both elegance and complexity in their design, guiding cell function through a balance of order and controlled disorder, underpinned by core thermodynamic laws. As we continue to unravel these molecular intricacies, the interplay between thermodynamics and biological function remains central in deciphering the essence of life.

11.5 Binding and Allosteric Mechanisms

The study of binding and allosteric mechanisms is foundational to understanding how proteins, particularly enzymes and receptors, regulate biological functions. These processes involve the interaction of proteins with ligands through specific sites, influencing enzymatic activity and signaling pathways. Allosteric regulation, a form of protein modulation involving conformational changes often triggered by ligand binding at sites distinct from active sites, plays a crucial role in ensuring finely tuned biological responses.

In discussing binding mechanisms, a ligand's affinity for a protein can be evaluated using the dissociation constant (K_d), a crucial parameter that quantifies the ligand concentration required to occupy half of the binding sites at equilibrium. Mathematically, the binding equilibrium between a protein (P) and ligand (L) is represented as:

$$P + L \rightleftharpoons PL$$

The equilibrium dissociation constant (K_d) is expressed as:

$$K_d = \frac{[P][L]}{[PL]}$$

A lower K_d value indicates a higher affinity, manifesting stronger binding interactions, while a higher K_d signals weaker binding. Several factors influence these interactions, including hydrogen bonding, van der Waals forces, electrostatic interactions, and hydrophobic effects. Entropic and enthalpic components are tightly associated with binding affinity; enthalpic contributions arise from interaction energies, and entropic factors relate to conformational and solvation changes upon binding.

Moving beyond simple binding interactions, allosteric regulation involves the modulation of protein activity through conformational changes induced by allosteric effectors binding to an allosteric site, separate from the active site. This binding induces a shift in the protein's energy landscape, altering the active site's configuration and thereby modulating activity. Allosteric mechanisms significantly contribute to enzymatic regulation, enabling the cell to maintain homeostasis, adapt to changes, and efficiently utilize resources.

The allosteric effect is often described by the concerted and sequential models of allostery. The concerted model, also known as the Monod-Wyman-Changeux (MWC) model, postulates that all subunits of an oligomeric protein transition between states in a concerted manner. Proteins can exist in the tense (T) state, with low affinity for substrate, and a relaxed (R) state, with high affinity. Binding of substrate or an effector stabilizes the protein in the R state, enhancing substrate affinity.

Conversely, the sequential model, or Koshland-Némethy-Filmer (KNF) model, proposes that substrate binding occurs in a stepwise fashion, with binding of a substrate to one site inducing conformational changes selectively transmitted to neighboring subunits, facilitating subsequent substrate binding. This model captures the subtlety of allosteric transitions and the partial conformational changes observed in many allosteric proteins.

An exemplar allosteric protein, hemoglobin, demonstrates the concerted model beautifully. Hemoglobin transitions between the T and R states modulated by oxygen binding. As oxygen binds, hemoglobin's affinity for additional oxygen molecules increases via cooperative binding, evidenced by the sigmoid shape of its oxygen dissociation curve, a hallmark of allosterism. Such behavior contrasts with the hyperbolic binding curve of myoglobin, which lacks allosteric regulation.

Within metabolism, allosteric enzymes like phosphofructokinase-1 (PFK-1) integrate multiple allosteric signals, allowing metabolic pathways to respond dynamically to cellular needs. PFK-1, a key regulatory enzyme in glycolysis, is activated by AMP and fructose-2,6-bisphosphate and inhibited by ATP and citrate, demonstrating an intricate balance between energy supply and demand.

Protein kinases offer another prime example of allosteric regulation, wherein their activation is controlled by substrate and cofactor binding as well as intramolecular phosphorylation events. Cyclin-dependent kinases (CDKs), pivotal regulators of the cell cycle, are activated upon binding to cyclins. Cyclin binding induces allosteric conformational changes enhancing their kinase activity, ensuring precise cell cycle control.

Allosteric inhibitors and activators have immense therapeutic potential, offering opportunities for drug design by targeting allosteric sites as opposed to active sites. Allosteric modulators possess the advantage of greater specificity and reduced likelihood of developing resistance compared to active site-directed inhibitors. The development of allosteric modulators, such as the BCR-Abl inhibitors in chronic myeloid leukemia treatment and G-protein coupled receptor (GPCR) modulators, marks significant strides in precision medicine.

Moreover, allosteric regulation extends beyond enzymes to encompass receptors and transcription factors, where signaling pathways are modulated by ligand-induced conformational changes. The nuclear receptors, for instance, mediate transcriptional responses to ligands, involving ligand binding to modulate allosteric conformations crucial for binding to DNA and coactivators.

The thermodynamics of binding and allosteric processes also underscore the integral role of entropy and enthalpy in conformational dynamics. Changes in Gibbs free energy (ΔG) during protein-ligand interactions are influenced by both ΔH and $-T\Delta S$, where conformational adjustments upon binding may either stabilize or destabilize protein conformations through entropic effects, such as the release of structured water molecules or structural flexibility alterations.

Complementary to mechanistic models, advances in structural biology, including cryo-electron microscopy and nuclear magnetic resonance, provide atomistic insights into allosteric transitions and binding site dynamics, enhancing our understanding of protein conformational landscapes and guiding rational drug design efforts.

Binding and allosteric mechanisms embody fundamental principles of protein dynamics essential to biological regulation. By modulating protein function through specific ligand interactions, these processes orchestrate the complex molecular symphony that underlies cellular and physiological regulation, offering profound implications in medicine and biotechnology. Through continued exploration, the field seeks not only to expand our fundamental understanding but also to harness the inherent potential of these mechanisms in innovation and therapeutic development.

11.6 Thermodynamics of Membrane Transport

The thermodynamics of membrane transport plays a critical role in understanding how substances move across biological membranes, a process essential for maintaining cellular homeostasis and facilitating communication with the external environment. Membrane transport involves a range of mechanisms, including passive and active transport, each dictated by the principles of thermodynamics. This section explores the energetic considerations of these processes, analyzing how living cells harness energy gradients to regulate their internal environments.

At the fundamental level, membrane transport can be divided into passive and active processes. Passive transport involves the movement of molecules down their concentration gradient without the input of external energy, relying instead on the inherent kinetic energy of molecules and the thermodynamic principle of diffusion. Active transport, in contrast, requires energy to move molecules against their concentration gradient, typically supplied by adenosine triphosphate (ATP) or ion gradients.

The driving force for passive transport is the Gibbs free energy difference across the membrane, which can be expressed by the relation:

$$\Delta G = RT \ln \left(\frac{C_{\text{in}}}{C_{\text{out}}} \right)$$

where R is the universal gas constant, T is the temperature in Kelvin, and C_{in} and C_{out} represent the concentration inside and outside the membrane, respec-

11.6. THERMODYNAMICS OF MEMBRANE TRANSPORT

tively. In a scenario where ΔG is negative, transport occurs spontaneously, facilitating equilibrium between the two sides of the membrane.

Passive transport encompasses several subtypes, including simple diffusion, facilitated diffusion via channel proteins, and carrier-mediated diffusion. Simple diffusion allows nonpolar molecules and small, uncharged polar molecules to traverse the lipid bilayer, driven by concentration gradients. Facilitated diffusion, however, involves transporters or channels that aid in the movement of ions and larger polar molecules. Aquaporins, for example, are channel proteins that facilitate the rapid transport of water molecules across the cell membrane, maintaining osmotic balance.

Ion channels are another critical component of facilitated diffusion. These transmembrane proteins form pores that allow specific ions to flow down their electrochemical gradients. The solute movement through channels like voltage-gated, ligand-gated, or mechanically gated channels is a fundamental process in signalling and homeostatic functions, crucial in generating action potentials within neurons.

Active transport, in contrast, necessitates a direct input of energy to move solutes against their concentration or electrochemical gradients. Contrary to passive transport, it is characterized by a positive Gibbs free energy change, necessitating coupling with an energy source. The classic example of primary active transport is the sodium-potassium pump (Na^+/K^+ ATPase), which maintains the Na^+ and K^+ gradient crucial for various cellular functions including volume regulation and membrane potential establishment. The energy from ATP hydrolysis alters the conformation of the pump, allowing ions to translocate across the membrane:

$$3Na^+_{in} + 2K^+_{out} + ATP \rightarrow 3Na^+_{out} + 2K^+_{in} + ADP + P_i$$

Secondary active transport, or co-transport, leverages the energy stored in the form of ion gradients established by primary active transport to drive the movement of other solutes. Symporters and antiporters, or exchangers, facilitate this process. The sodium-glucose transporter, for instance, uses the sodium gradient established by Na^+/K^+ ATPase to co-transport glucose into cells against its concentration gradient, illustrating energy coupling in secondary active transport.

Other transport processes driven by membrane potential gradients include electrogenic pumps that contribute to membrane potential formation. These pumps allow cells to store energy in the form of electrochemical gradients, which can be harnessed for subsequent transport processes or to conduct electrical impulses.

Membrane transport thermodynamics also extends into physiological phenomena like osmoregulation, where solute and water transport across membranes maintain cellular turgor and volume. Osmosis can be understood thermodynamically in terms of free energy changes associated with solvent movement, aided by water channel proteins and influenced by solute concentrations.

Additionally, the lipid composition of membranes affects transport properties, as fluidity and permeability are influenced by factors like cholesterol content, fatty acid saturation, and temperature. These lipid characteristics influence membrane thermodynamics, affecting transport kinetics and protein function.

Dual concerns of selectivity and transport rate are inherent in the evolutionary optimization of membrane proteins, reflecting a balance between efficiency, specificity, and energy cost. These proteins demonstrate intricate structural features, such as gating mechanisms in ion channels, conformational flexibility in transporters, and substrate binding affinities, underlining sophisticated regulatory systems that cells evolved to meet functional demands.

Biotechnological advancements leverage our understanding of membrane transport thermodynamics, aiming to design drugs that target transport systems, modulate ion channel activities, or correct malfunctioning transporters. Insights from these endeavors continue to enrich pharmacological strategies, particularly for conditions like hypertension, cystic fibrosis, and neurological disorders.

Through a cooperative regulation of passive and active processes, cells tactically maintain homeostasis, channeling dynamic energies to fulfill life's essential biochemical and physiological functions. This sophisticated orchestration embodies the living cell as a consummate energy manager, constantly navigating the delicate balance between structure, function, and environmental fluctuations.

11.7 Applications of Thermodynamics in Biotechnology

The application of thermodynamics to biotechnology provides a fundamental scientific basis for designing and optimizing biological processes aimed at producing products and solutions critical to medicine, agriculture, and industrial manufacturing. This section delves into the multifaceted applications of thermodynamics in biotechnology, examining how thermodynamic principles guide the development of processes ranging from drug design to bioengineering, and from metabolic pathway optimization to green technology.

In drug design and development, thermodynamics plays a pivotal role in understanding and optimizing ligand-receptor interactions. The efficacy of drugs largely depends on their ability to bind selectively and effectively to biological targets, such as enzymes or receptors. Thermodynamic parameters, such as changes in Gibbs free energy (ΔG), enthalpy (ΔH), and entropy (ΔS), provide insights into binding affinity and specificity. The Gibbs free energy change for binding can be expressed as:

$$\Delta G = \Delta H - T\Delta S$$

Here, a more negative ΔG indicates stronger binding affinity of a drug to its target, correlating with increased efficacy. Understanding the enthalpic and entropic contributions enables the optimization of binding interactions, balancing favorable enthalpic interactions (such as hydrogen bonding and van der Waals forces) with entropic considerations, including solvation effects and conformational flexibility. These aspects are crucial in the rational drug design process, where binding affinity and specificity are optimized through quantitative structure-activity relationship (QSAR) models that incorporate thermodynamic data.

In the field of enzyme engineering, thermodynamics informs the re-design of enzymes to enhance catalytic efficiency or to modify substrate specificity. Through directed evolution or rational design, enzymes are optimized to function under industrial conditions, such as extreme temperatures or pH levels, often beyond the range of their native environment. Assessing the activation energy and thermodynamic stability of enzyme-substrate complexes is critical for achieving desired efficiency and resilience. For instance, enhancing thermal stability of industrial enzymes used in the biofuel industry by engineering disulfide bridges or boosting hydrophobic core packing ensures robust performance.

Metabolic engineering, the practice of optimizing genetic and regulatory processes within cells to increase the production of a desired substance, heavily relies on thermodynamic principles to predict and control flux through metabolic networks. Utilizing metabolic control analysis (MCA) and flux balance analysis (FBA), engineers can simulate the impact of altering enzyme levels or knockout/overexpressing genes. Thermodynamic feasibility and energy balance calculations help predict pathway yields, guiding the redesign of metabolic pathways to produce pharmaceuticals, biofuels, or specialty chemicals with higher efficiency.

Bioprocess optimization for the production of proteins, vaccines, or other bioproducts benefits from thermodynamic principles guiding fermentation conditions. Parameters such as temperature, pH, and substrate concentrations are optimized for maximal yield and reduced energy consumption. The Gibbs free energy change associated with these bioprocesses determines their spontaneity and efficiency, often requiring careful engineering of microbial strains to ensure high product yields with minimal byproduct formation.

In system biology, thermodynamic modeling enables the investigation of complex networks of biomolecular interactions within cells. Such models incorporate thermodynamics to simulate the dynamic behavior of metabolic and signaling pathways, providing insights into cellular responses to environmental changes. These models help identify key regulatory nodes and potential intervention points, improving our understanding of diseases and informing therapeutic strategy development.

Thermodynamics also underpins the development of biosensors, which employ bioreceptors such as enzymes, antibodies, or nucleic acids for detecting analytes of interest, from metabolites to pathogens. The binding events in biosensors are governed by thermodynamic parameters that define the specificity, sensitivity, and response time. Optimizing these thermodynamic factors directly impacts sensor performance, crucial for clinical diagnostics and environmental monitoring.

In sustainable biotechnology, principles of green chemistry are aligned with thermodynamic concepts to design processes that minimize energy use and environmental impact. Biocatalysis, utilizing enzymes for chemical reactions, offers a more thermodynamically favorable and environmentally benign alternative to traditional chemical processes, reducing the need for harsh chemicals

11.7. APPLICATIONS OF THERMODYNAMICS IN BIOTECHNOLOGY

or conditions. Furthermore, thermodynamic analyses enable the assessment of carbon footprints and energy efficiency in biomass conversion technologies, promoting the development of sustainable and renewable energy sources.

Finally, the application of thermodynamics extends to material science, informing the synthesis of biomaterials with desired mechanical and chemical properties. By understanding the energetic contributions of molecule-molecule and molecule-surface interactions, scientists design biocompatible materials for applications ranging from surgical implants to tissue engineering scaffolds.

Thermodynamic applications in biotechnology provide a rigorous framework for examining biological processes, driving innovation across diverse fields. By harmonizing biological principles with thermodynamic insights, biotechnology achieves heightened precision and efficiency, paving the way for advanced therapeutic, industrial, and environmental solutions that align with the pressing demands of contemporary society.

Chapter 12

Advanced Topics in Physical Chemistry

Advanced topics in physical chemistry delve into cutting-edge research and innovative methodologies that expand the boundaries of chemical understanding. This chapter navigates through the complexities of computational chemistry, highlighting simulation techniques that model molecular interactions with unprecedented precision. The exploration of nonequilibrium thermodynamics challenges traditional views by addressing systems far from equilibrium, while supramolecular chemistry reveals the delicate architectures built from non-covalent interactions. Further insights are provided by molecular reaction dynamics, which detail the pathways and mechanisms at play during chemical transformations. As the chapter progresses, it examines the implications of emerging technologies, such as quantum computing and green chemistry, which propel the discipline into novel realms of inquiry and application.

12.1 Computational Chemistry and Simulation Techniques

Computational chemistry and simulation techniques represent an essential domain in contemporary physical chemistry, enabling the modeling and understanding of chemical systems with remarkable accuracy. Within this section, we shall delve into the fundamental methodologies and tools employed, as well as their applications in predicting molecular properties and dynamics. This exploration will enhance the reader's understanding of the crucial role computation plays in the study of chemical phenomena.

The crux of computational chemistry lies in the ability to solve the Schrödinger equation for complex systems. For isolated atoms and simple molecules, analytical solutions may suffice, but as the complexity and size of the system increase, approximations and numerical solutions become indispensable. At the heart of computational chemistry are quantum mechanical methods that simulate electronic structure. The methods vary widely in complexity and computational demand, ranging from ab initio methods, such as Hartree-Fock (HF) and post-Hartree-Fock methods (e.g., Møller-Plesset perturbation theory and Coupled Cluster methods), to semi-empirical and density functional theory (DFT) approaches.

The Hartree-Fock method represents a significant development in electronic structure theory. It involves an approximation wherein many-electron wave functions are constructed as Slater determinants of one-electron functions, known as orbitals. In this approach, electron-electron interactions are averaged, leading to significant computational simplifications. While HF theory serves as a cornerstone for more advanced methods, its limitations due to electron correlation effects have fostered the development of post-Hartree-Fock techniques.

Among these advanced techniques, Møller-Plesset perturbation theory (MP2) and the Coupled Cluster (CC) method stand out. MP2 provides a computationally feasible correction to the HF energy by considering pairwise electron correlation through a second-order perturbation approach. CC methods, on the other hand, include higher-order electron correlation effects, granting chemists the ability to achieve highly precise results for various chemical systems. Despite their accuracy, such methods are often limited by their computational intensity, particularly for large molecular systems.

Density Functional Theory (DFT) introduces a different paradigm wherein the electron density, rather than the wave function, serves as the fundamental quantity. DFT offers a balance between computational efficiency and accuracy, making it the method of choice for large-scale electronic structure calculations. The Kohn-Sham formulation of DFT elegantly simplifies the many-body problem by introducing a system of non-interacting electrons generating

12.1. COMPUTATIONAL CHEMISTRY AND SIMULATION TECHNIQUES

the same density as the real interacting system. Various exchange-correlation functionals have been developed to enhance the predictive power of DFT, catalyzing its application across numerous fields in chemistry and materials science.

In addition to electronic structure methods, molecular mechanics (MM) plays a pivotal role in the simulation of large biomolecules and complex molecular systems. MM treats molecules as collections of atoms connected by bonds, using classical mechanics to describe the potential energy surfaces. Force fields, such as AMBER, CHARMM, and OPLS, define the functional forms and parameters for the potential energy of molecular interactions, including bond stretching, angle bending, torsion, and non-bonded interactions like van der Waals forces and electrostatics. Force field development continues to be an active area of research to capture the intricacies of molecular interactions more accurately.

The coupled use of molecular mechanics and quantum mechanical approaches is exemplified in the QM/MM method, which provides an efficient way to simulate large systems by treating a small part of the system quantum mechanically, while the remainder is modeled using classical mechanics. Such an approach is particularly useful in studying enzymatic reactions and other processes where chemical reactivity is localized.

Beyond static calculations, molecular dynamics (MD) simulations represent another class of indispensable computational tools. MD enables the exploration of the temporal evolution of molecular systems by numerically integrating the classical equations of motion. Through MD simulations, researchers obtain insights into conformational dynamics, transport phenomena, and thermodynamic properties. The outcomes of MD simulations provide vital connections between microscopic interactions and macroscopic observables. Enhanced sampling techniques, such as replica exchange and metadynamics, have been developed to overcome the challenge of limited phase space sampling in MD simulations.

The realm of computational chemistry extends beyond the quantum and classical treatment of molecular systems, encompassing statistical mechanics and machine learning techniques. Statistical mechanics provides a bridge by connecting molecular-scale interactions with thermodynamic quantities, facilitating the derivation of ensemble properties and phase behaviors. Machine learning, with its rapidly expanding capabilities, offers novel approaches for predicting molecular properties, optimizing materials design, and elucidating complex reaction pathways. These data-driven methods complement traditional quantum chemistry techniques by providing faster predictions and enabling the exploration of vast chemical spaces.

A vital aspect of computational chemistry is the visualization and analysis of results, which facilitates the interpretation of data and validation of models. Visualization tools allow researchers to interact with three-dimensional molecular structures, observe dynamic processes, and identify key interactions. To verify and validate computational results, extensive benchmarking against experimental data is necessary. This synergy between simulation and experimentation is crucial for advancing chemical understanding and developing new materials and technologies.

In practical applications, computational chemistry aids in drug design, materials science, and environmental studies, among other fields. In drug discovery, computational methods allow for the screening of large libraries of compounds, the prediction of binding affinities, and the simulation of protein-ligand interactions. In materials science, the design of novel catalysts, polymers, and nanomaterials benefits from the insights gained through computational simulations. Environmental studies leverage computational tools to understand atmospheric reactions, pollutant behavior, and potential remediation strategies.

The future of computational chemistry and simulation techniques is fortified by advancements in high-performance computing and algorithm development. Quantum computing emerges as a potential game-changer, promising to solve currently intractable problems by leveraging the principles of quantum mechanics. Despite its nascent stage, quantum computing could revolutionize the way chemists model and simulate complex systems.

Potential applications of quantum computing in computational chemistry include solving electronic structure problems with unprecedented speed and accuracy, exploring new reaction mechanisms, and optimizing materials with tailored properties. Moreover, as hybrid classical-quantum approaches develop, they may offer new paradigms for molecular simulations, extending beyond the capabilities of current classical techniques.

Computational chemistry and simulation techniques stand at the forefront of contemporary physical chemistry. By enabling accurate predictions of molecular properties and dynamics, these methods not only deepen our understanding of chemical systems but also empower interdisciplinary research addressing pressing scientific and societal challenges.

12.2 Advanced Quantum Chemistry Methods

The evolution of quantum chemistry has reached unprecedented heights with the development of advanced methods that enable chemists to gain deeper insight into the behavior of electrons in complex chemical systems. This section delves into sophisticated quantum chemistry techniques, focusing on post-Hartree-Fock methods and their essential applications in accurate chemical modeling. These methods offer improved approximations for addressing electron correlation, a critical factor in achieving precise descriptions of molecular energy and properties.

The Hartree-Fock (HF) method, through its employment of self-consistent field (SCF) theory, initiated a powerful framework for wave function calculation by approximating the many-electron wave function as a single Slater determinant. However, as a mean-field theory, HF calculations fall short in accounting for electron correlation effects beyond exchange interactions, necessitating the development of post-Hartree-Fock methods. These advancements aim to include electron correlation more explicitly, refining the computational predictions of electronic structures and properties.

Among the widely used post-Hartree-Fock methods, configuration interaction (CI) and coupled cluster (CC) theories stand out for their ability to accurately account for electron correlation. Configuration interaction, an approach that solves the full electronic Schrödinger equation by expanding the wave function into a linear combination of Slater determinants, is an exact treatment limited only by computational resources. Its truncated variant, CI singles and doubles (CISD), includes electron excitations up to double excitations, providing a practical compromise between computational feasibility and accuracy.

Coupled cluster theory, notably the coupled cluster singles and doubles (CCSD) and CCSD(T) — where the latter includes perturbative triples, has become the gold standard for electronic structure calculations. It effectively includes electron correlation through an exponential ansatz, offering size-consistent and accurately correlated energies for ground and excited states. The ability to systematically include higher-order excitations distinguishes CC theories in delivering reliable predictions of molecular properties.

Multireference methods, such as multireference configuration interaction (MRCI) and multireference perturbation theory (MRPT), extend the post-Hartree-Fock landscape, targeting systems with near-degenerate electronic states common in transition metals and excited states of molecules. These methods leverage a reference wave function typically derived from complete active space self-consistent field (CASSCF) calculations, which account for strong electron correlation within a predefined orbital space, further incorporating weak correlation effects in subsequent steps.

Density Functional Theory (DFT) continues to advance alongside wave function methods, given its unique approach in modeling electronic structure through electron density rather than wave functions. Within DFT, significant strides have been made with the development of hybrid functionals, such as B3LYP, combining exact exchange with exchange-correlation functionals derived from planetary approaches to improve accuracy. Range-separated functionals, incorporating long-range corrections, address shortcomings in the treatment of long-range electron correlation and van der Waals interactions.

Theoretical chemists often blend several methodologies to overcome individual limitations and harness complementary strengths for particular applications. The application of highly accurate quantum chemical methods extends beyond small molecules to larger, complex systems through hybrid techniques like QM/MM, where reactive regions are treated with quantum mechanics and surrounding environments with molecular mechanics.

To showcase post-Hartree-Fock method applications, we look into spectroscopic properties and reaction mechanisms. For instance, CC and CI methods have played significant roles in accurately determining spectra, elucidating fine details in nuclear magnetic resonance (NMR), electron spin resonance (ESR), and ultraviolet-visible (UV-Vis) spectra with high precision. Accurate predictions align closely with experimental observations, bolstering the credibility of these computations. Reaction mechanisms, especially transition state theory, benefit immensely from insights obtained through transition states and reaction path optimizations using these advanced techniques.

Quantum Monte Carlo (QMC) methodologies complement traditional approaches by providing high accuracy in energy calculations through stochastic sampling of electron configurations. Variational and diffusion Monte Carlo methods offer promising alternatives to deterministic wave function methods, especially in capturing electron correlation in extended systems and solid-state chemistry.

In addition to methodological advances, computational quantum chemistry has witnessed transformative changes due to increasing computational power and novel algorithms. The parallelization of computations and the implementation of linear-scaling techniques allow for the tackling of larger systems with post-Hartree-Fock accuracy. The advent of machine learning tools presents opportunities to develop predictive functions and optimize molecules at a fraction of traditional computational costs.

Applications of advanced quantum chemistry methods extend to a wide array of fields beyond spectroscopy and reaction dynamics, impacting materials science, drug discovery, and catalysis. In material science, predicting band structures, electron transport properties, and defect states in semiconductors demonstrate the profound influence of accurate electronic structure predictions. Drug discovery and pharmacology harness these methods to optimize ligand-receptor interactions, facilitating the design of potent pharmaceuticals. Catalysis, a field where quantum chemistry intersects with chemical engineering, benefits from mechanistic insights into catalytic cycles, informing the design of efficient catalysts with industrial relevance.

Future developments in quantum chemistry are poised to address persistent challenges, such as strongly correlated systems and the development of universally applicable, accurate, and efficient functionals in DFT. Quantum computing emerges as a potential revolution, offering algorithms like quantum phase estimation and variational quantum eigensolver (VQE) that promise exponential speed-ups for electronic structure problems. As this landscape evolves, hybrid classical-quantum algorithms are anticipated to play pivotal roles in bridging the current technological gap.

Advanced quantum chemistry methods remain pivotal in the accurate modeling of chemical systems. The growing interplay among different methodologies and technological advances continues to enhance our capability to simulate systems of increasing complexity. The expansion of quantum chemical knowledge not only enriches our fundamental understanding of chemical processes but also propels various scientific domains forward, fostering innovation and technological advancement.

12.3 Nonequilibrium Thermodynamics

Nonequilibrium thermodynamics is a branch of physical chemistry that extends beyond the classical equilibrium paradigm to describe systems far from equilibrium. Traditional thermodynamics primarily deals with systems in or near a state of equilibrium, where properties are well-defined and state variables are constant over time. However, most natural and industrial processes occur away from equilibrium, requiring an approach capable of capturing time-dependent behaviors and spatial variations.

The foundational idea of nonequilibrium thermodynamics is to describe the transport processes and rate phenomena that arise due to gradients in temperature, pressure, chemical potential, and other potentials. These processes include heat conduction, diffusion, chemical reactions, and external forces acting on systems. The complexity of nonequilibrium thermodynamics arises from the need to predict the evolution of macroscopic properties as systems draw closer to or further from equilibrium.

The framework of nonequilibrium thermodynamics was fortified by the pioneering work of scientists like Onsager, Prigogine, and others. Key elements of this framework include the concepts of fluxes and forces, linear irreversible thermodynamics, and the Onsager reciprocal relations. These elements form the basis for understanding the transport processes in open systems and serve as a bridge between thermodynamics, statistical mechanics, and kinetics.

At the heart of nonequilibrium thermodynamics is the idea of entropy production. In contrast to equilibrium systems, where entropy remains constant or reaches a maximum, nonequilibrium systems experience a continuous production of entropy. The entropy production rate can be linked to irreversible processes occurring within the system, providing insight into the efficiency and directionality of these processes.

The basic form of the entropy production rate equation for a closed system is expressed as

$$\frac{dS}{dt} = \frac{d_e S}{dt} + \frac{d_i S}{dt}$$

where $\frac{dS}{dt}$ is the change in entropy over time, $\frac{d_e S}{dt}$ represents the change due to exchange with the environment, and $\frac{d_i S}{dt}$ is the internal entropy production, always non-negative according to the second law of thermodynamics.

For open systems, which exchange matter and energy with their surroundings, nonequilibrium thermodynamics introduces fluxes and thermodynamic forces. A flux is a measure of the flow of energy or matter per unit area per unit time, such as heat flux or matter flux. Thermodynamic forces, also known as affinities, are the gradients that drive these fluxes, such as temperature gradients for heat transport or concentration gradients for diffusion.

The linear relationship between fluxes and forces, valid under the assumption

12.3. NONEQUILIBRIUM THERMODYNAMICS

of local equilibrium, is expressed through phenomenological equations. The simplest form of these relationships is given by

$$J_i = \sum_j L_{ij} X_j$$

where J_i denotes the flux of the i-th quantity, L_{ij} are the phenomenological coefficients, and X_j are the thermodynamic forces or affinities. The Onsager reciprocal relations impose symmetry conditions, such that $L_{ij} = L_{ji}$.

These equations provide a powerful method to predict the behavior of systems undergoing transport processes. For example, in a simple heat conduction scenario, Fourier's law of heat conduction relates the heat flux J_q to the temperature gradient $-\nabla T$ by

$$J_q = -\kappa \nabla T$$

where κ is the thermal conductivity, a measure of the material's ability to conduct heat. Similarly, Fick's first law of diffusion relates matter flux to concentration gradients.

However, many systems of interest involve non-linearities or are driven by complex interactions between fluxes and forces. In these situations, nonlinear nonequilibrium thermodynamics must be applied, which considers more comprehensive phenomenology and often requires numerical simulations for accurate descriptions.

One significant achievement in the field is the discovery of non-equilibrium steady states, where systems maintain constant fluxes and forces over time despite being far from equilibrium. These states reflect a delicate balance between the influx and outflux of energy and matter, leading to constant entropy production rates. Non-equilibrium steady states form the backbone of many biological processes and industrial operations.

For instance, living cells exemplify highly orchestrated nonequilibrium systems, where metabolic processes maintain concentration gradients across membranes, powering essential biological functions. The coupled transport of ions and molecules across cellular membranes serves as the basis for nerve impulse transmission, energy conversion, and homeostasis.

Another area of interest is the study of dissipative structures, championed by Prigogine, where ordered structures emerge spontaneously from initially disordered nonequilibrium systems. Dissipative structures represent regions of decreased entropy production and occur when positive feedback mechanisms promote specific patterns, such as convection rolls in fluid dynamics and pattern formation in chemical reactions like the Belousov-Zhabotinsky reaction.

Nonequilibrium thermodynamics also guides the analysis of chemical reaction kinetics, particularly important in catalysis and enzymatic processes, where reactions proceed under nonequilibrium conditions. Detailed balance, a principle stating that at equilibrium, each microscopic process and its reverse occur at equal rates, can be broken in nonequilibrium conditions, leading to net chemical transformations.

Within the industrial context, nonequilibrium thermodynamics is applied in synergetic processes involving heat, mass, and momentum transfer. For example, energy generation in power plants, chemical production in reactors, and refrigeration systems frequently operate under nonequilibrium conditions, necessitating consideration of energy efficiency and system stability. The quest for increased efficiency in energy conversion technologies often revolves around optimizing nonequilibrium processes and reducing entropy production.

In advanced materials research, nonequilibrium thermodynamic principles help design nonequilibrium processing techniques such as rapid solidification, which can yield materials with enhanced mechanical properties. The processing of polymers and nanomaterials frequently involves nonequilibrium thermodynamics, as does the development of smart materials capable of adaptive responses to environmental stimuli.

The pursuit of sustainable technologies and green chemistry also intersects with nonequilibrium thermodynamics, as many environmental processes, including atmospheric reactions, combustion, and pollutant degradation, are inherently nonequilibrium. Modeling these complex processes enables the design of mitigation strategies and informs policy decisions.

In summary, nonequilibrium thermodynamics encompasses a rich tapestry of principles and applications that extend the classical scope of thermodynamics to embrace the dynamic complexity observed in the real world. By understanding the driving forces and behavior of systems away from equilibrium, scientists and engineers can harness these processes' potential in a multitude of scientific and technological advancements.

12.4 Soft Matter and Complex Fluids

Soft matter and complex fluids encompass a fascinating class of materials—characterized by their unique mechanical properties, responsiveness to external stimuli, and structural organization—that bridge the gap between solids and liquids. This area of study merges physics, chemistry, and materials science, exploring the intricate behaviors exhibited by polymers, colloids, liquid crystals, gels, foams, biological tissues, and other related systems.

The underlying hallmark of soft matter is the prevalence of mesoscopic structures, where characteristic lengths are larger than atomic scales but smaller than macroscopic dimensions. These systems are governed by weak interaction forces like van der Waals forces, hydrogen bonds, and electrostatic interactions, which allow significant molecular mobility and facilitate structural reconfiguration under modest conditions. Consequently, soft matter demonstrates distinctive properties such as viscoelasticity, self-assembly, and complex flow behaviors.

Polymers, long-chain molecules composed of repeating monomer units, exemplify the variety and richness of soft matter physics. The architecture of polymers can range from linear to branched and network structures, dictating their behavior and applications. Chain flexibility gives rise to phenomena like reptation, where polymer chains slide through entanglements in a snake-like motion. This results in viscoelastic behavior, where materials exhibit attributes of both viscous liquids and elastic solids, crucial to the processing and use of polymers in everyday applications.

Understanding polymer dynamics is essential for tailoring performance properties like toughness, elasticity, and adhesion, leveraged in industries ranging from packaging to biomedical devices. Advanced techniques such as neutron scattering, nuclear magnetic resonance (NMR), and rheometry are indispensable for probing these dynamics and unraveling the impact of structural parameters on macroscopic properties.

Colloids, another quintessential category within soft matter, are systems wherein microscopic particles are dispersed in a continuous medium. With sizes typically ranging from nanometers to micrometers, colloids present diverse interactions due to surface charges, steric hindrance, and depletion forces. These interactions govern colloidal stability, impacting aggregation and phase separation phenomena.

The phase behavior of colloids is notably captivating, as steric or electrostatic interactions can lead to crystallization, gelation, or glass formation. Such colloidal assemblies are visible in products like paints, cosmetics, and foodstuffs. Techniques like light scattering, microscopy, and optical tweezers enrich the understanding of these systems, enabling the design of stable colloids with tailored functionalities.

Liquid crystals, materials exhibiting ordered states between solid crystals and isotropic liquids, are valued for their ability to respond to external cues like temperature, electric fields, and mechanical stresses. They manifest unique optical properties, underpinning the technology of liquid crystal displays (LCDs), where electric field application induces reorientation of molecular director fields, varying the passage of light.

The study of liquid crystals probes the rich phase transitions among nematic, smectic, and cholesteric phases, each distinguished by its molecular arrangement and order. These phases provide insight into fundamental aspects of soft matter elasticity, defect structures, and hydrodynamics. The application potential spans beyond displays into areas such as smart windows, sensors, and drug delivery systems.

Gels, loosely cross-linked polymer networks swollen by a solvent, display characteristics of both solid-like and liquid-like materials. Their mechanical elasticity, highly tunable via cross-link density and solvent interactions, renders them suitable for diverse applications, from contact lenses and wound dressings to soft robotics. Responsive gels, which swell or contract in response to stimuli such as pH, temperature, and ionic strength, open new avenues in adaptive and biomimetic materials.

Foams, dispersions of gas in a liquid or solid matrix, serve as a paradigm for studying complex fluid mechanics and interfacial phenomena. Their stability hinges on the interplay of film elasticity, drainage, and coalescence dynamics. The transition from wet to dry foams elucidates percolation and jamming phenomena applicable in food science, fire suppression, and construction materials.

In exploring soft matter and complex fluids, the concept of self-assembly emerges as a unifying theme, driving the spontaneous organization of systems into higher-order structures. Self-assembly exploits weak interactions to create functional materials—illustrated by the formation of micelles, vesicles, and block copolymer structures—enabling the manufacture of nanostructured materials with precision and efficiency.

Biological systems, epitomes of soft matter, present a biomimetic inspiration by demonstrating hierarchical self-assembly and responsive functionalities—ranging from the structural integrity of cellular membranes to the mechanical properties of tissues. The integration of synthetic polymers with biological macromolecules catalyzes advancements in tissue engineering, drug delivery, and biocompatible devices.

The flow behavior of complex fluids, marked by their non-Newtonian characteristics, challenges traditional hydrodynamics by presenting phenomena such as shear thinning, shear thickening, and thixotropy. Rheological studies uncover these behaviors, enabling the optimization of processes such as extrusion, mixing, and coating. Modeling and simulating complex fluid dynamics spur innovative approaches to novel processing techniques and product formulations.

On a broader scale, soft matter and complex fluids emphasize addressing industry-driven problems such as sustainability, additive manufacturing, and responsive material design. Advanced computational tools, ranging from molecular dynamics simulations to continuum modeling, offer deeper quantitative insight into these systems, aiding in crafting materials with specific properties and functions.

The domain of soft matter and complex fluids encapsulates an array of materials distinguished by their adaptability, multifunctionality, and potential for innovation. These systems, united by their structural and dynamic complexity, provide an expansive playground for understanding fundamental scientific principles and guiding technology's future applications across multidisciplinary fields. Through the interplay of experimental techniques and theoretical frameworks, the study of soft matter deepens our appreciation of the delicate balance between structure, dynamics, and function.

12.5 Molecular Reaction Dynamics

Molecular reaction dynamics is a fundamental field of study in physical chemistry that investigates the detailed pathways and mechanisms of chemical reactions at the molecular level. Unlike traditional reaction kinetics, which mainly focuses on the rates and outcomes of reactions, molecular reaction dynamics delves into the motions and interactions of individual atoms and molecules during a chemical process. This exploration offers a comprehensive understanding of how chemical reactions unfold, providing insight into energy distributions, transitional states, and potential energy surfaces.

The primary objective of molecular reaction dynamics is to elucidate the step-by-step progression from reactants to products on a fundamental molecular scale. Central to this is the concept of the potential energy surface (PES), which represents the multi-dimensional landscape over which a chemical reaction occurs. Each point on the surface corresponds to a particular arrangement and energy of the participating atoms. Thus, studying the topography of the PES, including wells, barriers, and saddle points, is crucial for understanding reaction pathways and intermediates.

Molecular beam experiments have historically played a crucial role in advancing our understanding of molecular reaction dynamics. These experiments involve the production of a collimated beam of molecules that can be manipulated and observed to scrutinize the interactions under controlled conditions. By controlling the velocity, orientation, and internal state of the reactants, scientists can precisely dissect the reaction mechanism. This framework enables the study of phenomena such as collision dynamics, energy transfer, and molecular scattering in great detail.

A landmark achievement in this field is the cross-beam scattering experiment, where two molecular beams intersect at a known angle. Upon collision, products are formed and detected, allowing researchers to determine the angular distribution and energy partitioning among the products. This setup has been instrumental in confirming theoretical predictions and refining models of molecular reaction dynamics.

Theoretical approaches complement experimental techniques, employing advanced computational methods to simulate reaction dynamics. Classical trajectory simulations, quantum mechanical methods, and mixed quantum-classical approaches allow researchers to simulate the detailed motions and interactions of particles during chemical reactions. These simulations offer profound insights into the mechanisms of bond breaking and forming, energy redistribution, and the influence of quantum effects such as tunneling and zero-point energy.

One of the classic models in reaction dynamics is the transition state theory (TST), independently developed by Eyring and Polanyi in the early 20th century. TST postulates that a chemical reaction proceeds through a high-energy transition state, also known as the activated complex, situated at the saddle point of the potential energy surface. The theory explains reaction rates in terms of the population of this transient state and the frequency at which the system crosses the activation barrier. Though TST assumes equilibrium is maintained among various states during the reaction, corrections and extensions, including the variational transition state theory (VTST), account for more complex pathways and entropy considerations.

For reactions occurring in condensed phases or under the influence of solvents, the study of reactive dynamics extends beyond gas-phase models. Here, solvent effects can profoundly influence both the energetics and the dynamics of reactions. Solvent molecules can stabilize or destabilize reactants, intermediates, and transition states, and they can also participate in the reaction through solvation dynamics and providing alternate pathways. Dynamic solvent effects necessitate considering solvation shells, dielectric relaxation, and solute-solvent energy transfer mechanisms within the context of molecular dynamics simulations or implicit solvent models.

Chemical reaction dynamics is tightly intertwined with the field of spectroscopy, which facilitates direct observation of changes in molecular structure and energy levels. Ultrafast spectroscopy techniques, such as femtosecond laser spectroscopy, allow the capture of transient species and dynamics on the timescale over which they occur, often in the femtosecond to picosecond range. This capacity to "watch" chemical reactions in real-time offers an unprecedented window into transient intermediates and the evolution of electronic states during a reaction.

A notable development is in the study of reactive scattering processes, where detailed information on the energy and angular distribution of reaction products provides valuable clues to the underlying dynamics. For instance, in the hydrogen exchange reaction

$$H + D_2 \longrightarrow HD + D$$

experiments and simulations have yielded profound insights into vibrational energy disposal, reaction pathways, and the role of quantum resonances.

Reaction dynamics also contribute substantially to our understanding of catalysis, both homogeneous and heterogeneous. Catalysts accelerate chemical reactions by providing alternate pathways with lower activation energies, often involving complex surface interactions. On a molecular level, reaction dynamics help elucidate adsorption, desorption, diffusion, and surface reaction mechanisms critical to catalyst effectiveness. In heterogeneous catalysis, researchers utilize techniques such as surface-enhanced Raman spectroscopy (SERS) and scanning tunneling microscopy (STM) to gain molecular-level insights into the catalytic surfaces and active sites.

In biological systems, molecular reaction dynamics provide critical insights into enzymatic processes, signal transduction, and molecular recognition. Enzymes, often described as perfect catalysts, achieve remarkable rate enhancements through precise spatial organization and dynamics at their active sites. Studies in enzyme dynamics emphasize the role of conformational changes, induced fit, and dynamic allosteric regulation in modulating enzyme activity and specificity.

Reaction dynamics is instrumental in atmospheric chemistry, where the reactivity of radicals, ozone depletion, and pollutant formation are dependent on intricate molecular interactions and energy transfer processes. In environmental chemistry, understanding the dynamics of reactions involving greenhouse gases and pollutants contributes to developing strategies for environmental protection and sustainable practices.

The field continues to advance as new experimental techniques and computational methodologies emerge, driving deeper exploration into nonadiabatic effects, electronically excited states, and complex environmental conditions. Machine learning approaches are beginning to complement traditional methods by analyzing massive datasets to predict potential energy surfaces and reaction pathways with impressive accuracy.

The study of molecular reaction dynamics is central to a comprehensive understanding of chemistry, with implications spanning from fundamental scientific insights to practical applications such as drug discovery, materials development, and environmental sustainability. By unraveling the intricate dance of atoms and molecules during chemical transformation, molecular reaction dynamics provides the foundation for innovative approaches in science and technology.

12.6 Supramolecular Chemistry

Supramolecular chemistry delineates a dynamic frontier in chemical sciences, focusing on the study of entities formed through non-covalent interactions among molecules. This branch extends beyond classical molecular chemistry, exploring the vast array of structures and functions arising from molecular assemblies bound through weak interactions, such as hydrogen bonding, van der Waals forces, $\pi\pi$ interactions, electrostatic forces, and metal coordination.

12.6. SUPRAMOLECULAR CHEMISTRY

The advent of supramolecular chemistry has revolutionized our understanding of molecular functionality, driven by the principles of molecular recognition, self-assembly, and cooperation.

Central tenets of supramolecular chemistry rest on molecular recognition processes. Molecular recognition is the ability of one molecule (the host) to selectively bind a specific guest molecule through non-covalent interactions, akin to the lock and key model often observed in biological systems. This selectivity and specificity are paramount in designing molecular systems that mimic natural processes. Molecular recognition underpins numerous biological phenomena, including enzyme-substrate interactions, antigen-antibody binding, and receptor-ligand activities.

Crown ethers and cyclodextrins serve as exemplary models of molecules that exhibit highly selective recognition properties. Crown ethers, cyclic compounds containing several ether groups, selectively complex metal cations by size exclusion, resembling the selectivity seen in biological ion channels. Cyclodextrins, cyclic oligosaccharides, form inclusion complexes with various guest molecules through hydrophobic interactions, rendering them useful as solubility enhancers in pharmaceutical formulations.

The ambition to understand and replicate these biological processes has fostered the development of artificial receptors and sensors. These supramolecular systems, designed to mimic biological recognition, are invaluable in chemical sensing, where they are used to detect various analytes with high specificity. As an application, chemosensors for detecting ions or small organic molecules utilize fluorescence or colorimetric methods to signal the presence of the target analyte, providing tools crucial for environmental monitoring and medical diagnostics.

Self-assembly, another cornerstone of supramolecular chemistry, describes the spontaneous organization of components into ordered structures without external direction. This process is driven by the optimization of non-covalent interactions and often leads to the formation of diverse and complex architectures. The principles of self-assembly are deeply embedded in natural systems, such as protein folding, lipid bilayers, and DNA base pairing, showcasing nature's efficiency in creating structured environments under physiological conditions.

The design and synthesis of self-assembling systems have vast implications spanning nanotechnology, materials science, and drug delivery. One illustrative example is the assembly of amphiphilic molecules, which possess both hydrophilic and hydrophobic domains. These molecules self-organize into micelles, vesicles, and bilayers in aqueous environments, enabling encapsulation and controlled release of therapeutic agents, as seen in the case of liposome-based drug delivery systems.

Supramolecular polymers extend the concept of traditional polymers through reversible non-covalent bonding, allowing tunable mechanical and dynamic properties. These polymers exhibit stimuli-responsive behavior, altering their properties in response to changes in temperature, pH, or light, paving the way for smart materials used in coatings, adhesives, and biomedical devices.

Rotaxanes and catenanes epitomize mechanically interlocked molecular architectures in supramolecular chemistry. Rotaxanes, consisting of a dumbbell-shaped molecule threaded through a macrocyclic ring, and catenanes, formed by interlinked rings akin to a chain, demonstrate the intricate architectures possible through precision engineering of supramolecular systems. These molecular machines exhibit potential in storage, catalysis, and information processing, offering insights into mimicking mechanical motions at the nanoscale.

The burgeoning field of host-guest chemistry, a subset of supramolecular chemistry, illustrates complexation phenomena where one or more guest molecules bind within the cavity of a host. This encapsulation alters the chemical reactivity and physical properties of the guest, leading to responsive behaviors that have implications for catalysis and separation technologies. Calixarenes, cucurbiturils, and pillararenes are prototypical host molecules, acting as selective encapsulators for small molecules and ions.

Supramolecular assemblies are pivotal in the fabrication of novel nanostructured materials. Molecular self-assembly underpins the fabrication of supramolecular crystals, gels, and frameworks with precisely controlled porosities and functionalities. Metal-organic frameworks (MOFs) and covalent organic frameworks (COFs) are prime examples, extensively used for gas storage, separation, and catalysis applications. Their high surface areas and tunable properties make them candidates for capturing carbon dioxide, hydrogen storage, and pollutant removal.

Supramolecular chemistry also offers profound contributions to the development of stimuli-responsive materials, leading to "smart systems" capable of adapting their properties to environmental changes. This adaptability is employed in medical applications, such as responsive drug delivery systems, where drug release is triggered by specific physiological conditions.

The integration of supramolecular chemistry with nanotechnology and materials science is accelerating the innovation of electronic devices, with applications in molecular electronics and sensors. Electronic devices that harness supramolecular components—such as organic photovoltaic cells and light-emitting diodes—illustrate the potential of manipulating molecular interactions to harness electronic properties at the nanoscale.

Beyond technological applications, supramolecular chemistry holds implications in understanding biological processes by providing a molecular-level model to decipher complex biological systems. Molecular mimicry allows researchers to recreate biological structures and dynamics, assisting in drug design and understanding diseases. Supramolecular elucidation of protein folding and misfolding processes, for instance, sheds light on amyloid diseases and potential therapeutic strategies.

In summary, supramolecular chemistry, with its multifaceted interactions, self-assembly capabilities, and host-guest chemistry, is poised to transform scientific understanding and innovation. By harnessing the inherent properties of non-covalent interactions, chemists can design and create complex systems that imitate natural phenomena and significantly impact fields ranging from healthcare to environmental sustainability. Through continuous exploration, supramolecular chemistry continues to unlock new possibilities, driving an era of advanced functional materials and biologically inspired systems.

12.7 Emerging Topics and Future Directions

The field of physical chemistry is continually expanding, evolving to encompass novel technologies and methodologies that push the boundaries of our scientific understanding. Emerging topics and future directions in this area highlight transformative approaches that promise to redefine facets of chemistry and materials science, with profound implications for technology, environment, and health.

One of the most compelling emerging themes is the advancement in green chemistry and sustainable technologies. This global movement seeks to reformulate chemical processes and products to minimize hazardous substances, improve resource efficiency, and promote renewable feedstocks. The principles of green chemistry advocate for energy efficiency, waste reduction, and the use of safer solvents and conditions. For instance, catalysis plays a key role in enhancing reaction efficiency and selectivity, helping minimize byproducts and energy input.

The development of sustainable catalytic systems has seen significant progress, with photocatalysis and electrocatalysis offering cleaner alternatives for chemical reactions. These technologies harness light energy and electrical energy, respectively, to drive reactions under milder conditions. In photocatalysis, solar light is used to activate photocatalysts, facilitating processes such as water splitting and CO_2 reduction, integral to renewable energy strategies.

In renewable energy, artificial photosynthesis emerges as a promising field aiming to replicate natural photosynthesis to produce fuels from sunlight, water, and CO_2. This approach supports the transition towards solar fuels, leverages innovations in material science to develop efficient light-harvesting systems, and aims to address energy storage challenges associated with intermittent renewable sources. The integration of advanced materials such as metal-organic frameworks (MOFs) and semiconductor nanoparticles enhances the efficiency of these processes, offering a sustainable pathway for energy generation.

Nanotechnology continues to represent an area of immense potential within physical chemistry, with applications sprawling across disciplines. The unique properties of nanoscale materials, arising from quantum size effects and high surface-to-volume ratios, open avenues for innovations in drug delivery, imaging, sensing, and electronics. Advances in nanoscale synthesis techniques enable the precise control of material properties, producing nanoparticles with tailored functionalities.

In medicine, nanoparticles are engineered as delivery vehicles for therapeutics, capable of targeting specific cells or tissues, thus improving efficacy and reducing side effects. Quantum dots, with their size-tunable optical properties, are utilized in bioimaging, providing high-resolution and long-lived fluorescence markers. Beyond healthcare, nanostructured materials enhance sensor technologies, offering sensitivity and selectivity for detecting chemical and biological agents in various environments.

Quantum computing has garnered substantial attention as a revolutionary technology poised to solve complex problems unmanageable by classical computers. In physical chemistry, it enables the simulation of quantum systems with higher accuracy and less computational cost, potentially transforming fields like materials science, cryptography, and pharmacology. Algorithms designed for quantum computers, such as the Quantum Approxi-

12.7. EMERGING TOPICS AND FUTURE DIRECTIONS

mate Optimization Algorithm (QAOA) and Variational Quantum Eigensolver (VQE), promise breakthroughs in solving optimization problems and simulating molecular energies.

The convergence of artificial intelligence (AI) and machine learning with physical chemistry heralds a new paradigm in research and application. These technologies analyze vast datasets to predict molecular properties, identify reaction pathways, and optimize materials design. Machine learning algorithms, ranging from deep neural networks to ensemble methods, are employed to model complex systems, offering predictive power in understanding chemical phenomena, reducing the need for extensive experimental trials.

One practical implementation involves the acceleration of material discovery through predictive modeling, identifying promising candidates for catalysts, batteries, and polymers. AI-driven platforms facilitate the design of experiments, adaptive sampling, and real-time analysis, expediting the research and development cycle. As these digital tools mature, they offer personalized solutions, fostering innovation at unprecedented speeds.

Advanced analytical techniques continue to evolve, offering deeper insights into chemical systems at high spatial and temporal resolution. Techniques such as ultrafast spectroscopy, atomic force microscopy (AFM), and cryogenic electron microscopy (Cryo-EM) enable the *in situ* study of dynamic processes, unraveling intermediate states and transient species in chemical reactions. Such capabilities provide researchers with a fine-grained view of molecular interactions, enabling a detailed understanding of complex systems across disciplines from materials science to biological chemistry.

In parallel, the demand for interdisciplinary research intensifies, integrating principles from physics, biology, materials science, and engineering to address grand challenges in energy, health, and environment. Collaborative efforts facilitate novel solutions by combining diverse expertise and methodological approaches, catalyzing the development of hybrid technologies like biohybrid solar cells and biocompatible nanomaterials.

The driving forces behind these emerging topics and future directions include societal needs, environmental concerns, and technological aspirations. As global challenges like climate change, resource scarcity, and population growth reach critical levels, chemistry—and particularly physical chemistry—stands at the forefront of offering scientific solutions. The initiative for a circular economy, where resources are used more effectively and sustainably, finds its scientific backbone within these innovative advancements.

Furthermore, the education and training of the next generation of chemists become pivotal, necessitating an adaptive curriculum that encompasses the technological skills and interdisciplinary knowledge vital for tackling future challenges. In preparing scientists and engineers to navigate these landscapes, emphasis on skills in computational methods, data science, and ethical literacy will prepare them to contribute effectively to evolving scientific and industrial sectors.

Emerging topics and future directions in physical chemistry underscore a transition toward a more integrated and sustainable scientific practice. The convergence of cutting-edge technology, interdisciplinary collaboration, and ethical stewardship sets a dynamic course for physical chemistry, reflecting its critical role in advancing society and preserving the environment. Through these innovative trajectories, physical chemistry will continue to drive progress, addressing the needs and aspirations of a rapidly changing world.

12.7. EMERGING TOPICS AND FUTURE DIRECTIONS

www.ingramcontent.com/pod-product-compliance
Lightning Source LLC
Chambersburg PA
CBHW052142220526
45471CB00004B/1483